Strategies for Sustainability

For further volumes:
http://www.springer.com/series/8584

Strategies for Sustainability

Aims and Scope

The series, will focus on "implementation strategies and responses" to environmental problems – at the local, national, and global levels. Our objective is to encourage policy proposals and prescriptive thinking on topics such as: the management of sustainability (i.e. environment-development trade-offs), pollution prevention, clean technologies, multilateral treaty-making, harmonization of environmental standards, the role of scientific analysis in decision-making, the implementation of public-private partnerships for resource management, regulatory enforcement, and approaches to meeting inter-generational obligations regarding the management of common resources. We will favour trans-disciplinary perspectives and analyses grounded in careful, comparative studies of practice, demonstrations, or policy reforms. We will not be interested in further documentation of problems, prescriptive pieces that are not grounded in practice, or environmental studies. Philosophically, we will adopt an open-minded pragmatism – "show us what works and why" – rather than a particular bias toward a theory of the liberal state (i.e. "command-and-control") or a theory of markets.

We invite Authors to submit manuscripts that:
Prescribe how to do better at incorporating concerns about sustainability into public policy and private action.

Document what has and has not worked in practice.

Describe what should be tried next to promote greater sustainability in natural resource management, energy production, housing design and development, industrial reorganization, infrastructure planning, land use, and business strategy.

Develop implementation strategies and examine the effectiveness of specific sustainability strategies. Focus on trans-disciplinary analyses grounded in careful, comparative studies of practice or policy reform.

Provide an approach "…to meeting the needs of the present without compromising the ability of future generations to meet their own needs," and do this in a way that balances the goal of economic development with due consideration for environmental protection, social progress, and individual rights.

The Series Editors welcome any comments and suggestions for future volumes

SERIES EDITORS
Lawrence Susskind
susskind@mit.edu

Professor Ravi Jain
rjain@pacific.edu

Chittaranjan Ray • Ravi Jain
Editors

Drinking Water Treatment

Focusing on Appropriate Technology and Sustainability

Springer

Editors
Prof. Chittaranjan Ray
Civil & Environmental Engineering
University of Hawaii at Manoa
2540 Dole Street, Holmes Hall 383
Honolulu, HI 96822
USA
cray@hawaii.edu

Prof. Ravi Jain
School of Engineering & Computer Science
University of the Pacific
3601 Pacific Avenue
Stockton, CA 95211
USA
rjain@uop.edu

ISBN 978-94-007-1103-7 e-ISBN 978-94-007-1104-4
DOI 10.1007/978-94-007-1104-4
Springer Dordrecht Heidelberg London New York

Library of Congress Control Number: 2011930184

Cover design: deblik, Berlin

Printed on acid-free paper

Springer is part of Springer Science+Business Media (www.springer.com)

Preface

It is estimated that over 1.1 billion people do not have access to safe water (UNICEF Handbook on Water Quality, 2008). Clearly, this creates enormous human health and welfare challenges. The reasons for the unavailability of safe water relates to the enormous capital investment and operating expenses that must be incurred to be able to provide reliable and safe water; this is simply out of reach for most developing countries. This book was written to provide insight into the available sustainable technologies for producing an adequate safe water supply.

In many regions of the world, including the United States, rivers carry significant amounts of pollutants derived from industrial and municipal discharges, nonpoint sources such as agricultural and urban runoff, and accidental spillage. Water utilities that use surface water for supply must remove these chemicals in the plant prior to distribution. This involves the use of significant amounts of chemicals and advanced treatment technologies such as activated carbon or membrane units if micropollutants (e.g., pesticides, gasoline and solvent constituents) are present in the source waters. These technologies are expensive and they also need highly skilled operators.

Many small communities, even in industrialized countries, do not have such resources to meet the challenges. For long-term sustainability, incorporation of the most advanced technologies may not be feasible for small communities in developed countries and for most communities in developing countries. To respond to this crucial need, appropriate technologies are discussed in the book.

Water treatment methods such as solar distillation, solar pasteurization, membrane filtration utilizing techniques and materials that are affordable, and natural soil/aquifer filtration may be considered sustainable. These systems can function effectively at various scales and be able to provide potable water with very little need for additional treatment. Also, these technologies can be affordable in developing countries.

Solar distillation has been practiced in many arid and desert countries. In certain places, solar stills are coupled with membrane units for drinking water production. There are several variations of the stills used for drinking water production. One of the recent versions, patented by the US Department of Interior (inventor: J. Constantz), can be used for drip irrigating row crops and producing drinking water.

Solar pasteurization is one of the easiest methods to produce potable water in remote sunny areas. Heating water to a sufficiently high temperature for a certain time period destroys harmful microorganisms. It is also an inexpensive alternative in areas without electricity and water infrastructures. Common materials such as cylindrical plastic bottles can be used to pasteurize water by exposing the water to sunlight. A simple but effective method, with the tradeoff of low flow-rates.

Currently, membrane filtration is an expensive treatment technology and it is used for the desalination of sea water, brackish water, or other process waters. Depending on the pore sizes of the membranes, they are classified as "microfiltration," "ultrafiltration," "nanofiltration," and "reverse osmosis." Membrane cost and energy needed to pressurize the water chamber above the membranes control the per unit production cost of water. It is still possible to produce membrane filtrate from low-cost materials using alternate energy sources so that the process can be "democratized."

Natural filtration is a process that utilizes the pollutant adsorption and degradation capability of soil and aquifer materials and it has been formally deployed for drinking water production in Europe for more than a century. Wells, either vertical or horizontal, are placed some distance away from the river and are pumped on a sustained basis. This induces the river water to flow to the pumping wells. During soil and aquifer passage most contaminants from surface water are removed via sorption or degraded through microbial processes.

Biblical stories mention drinking water from a hole next to the Nile River rather than drinking the water from the river directly. In most areas of the developing world, especially in rural communities, the spread of cholera diminished after the use of hand pumps compared to the situation when surface water was used for drinking. Therefore, the soils and the underlying aquifer materials have tremendous capacity to remove surface water pollutants.

If properly designed and operated, most natural filtration systems (called bank filtration systems) do not need significant additional treatment with the exception of disinfection. However, excessive pumpage using infiltration galleries or scouring of riverbeds may reduce the effectiveness of such systems. In all instances, the quality of filtrate from these systems is still superior to that of the river water.

Provided in the book is a comparative analysis of drinking water treatment technologies that focus on appropriate technology and sustainability (Chap. 2). This chapter can serve as a means of comparing various sustainable treatment technologies for potential implementation. Some of the key technologies discussed are: *natural filtration, riverbank filtration, slow sand filtration, membrane filtration, solar pasteurization, membrane desalinization, and solar distillation.*

The chapter on transdisciplinary analysis provides information about sustainability concepts, industrial practices, sustainability of technology in developing countries, sustainability framework, and suggestions for technology transfer and implementation.

It is desirable to use less amounts of chemicals, energy, and manpower in drinking water production. Greater sustainability is achieved when comparable quality of water is produced without the need of excessive amounts of energy, labor, and expensive equipment/technology.

Some of the sustainability strategies that need to be examined in detail are: (a) reduction in chemical and energy use in water treatment, (b) production of water that contains less pathogens and disinfection byproducts compared to the use of surface water, and (c) focus on water utilities and communities (e.g., water treatment plants) to improve source water quality to reduce further treatment of the filtrate. If the watersheds are protected and the source water is of high quality, treatment technologies can be less costly and thus sustainable.

The authors are most grateful to the chapter contributors (as listed) and the reviewers who spent considerable time and effort to make this text possible. We are grateful to April Kam and Patricia Hirakawa (University of Hawaii), Kaben Kramer, and Deanna Henricksen (University of the Pacific) for their help with background research and for manuscript preparation.

Many individuals at Springer were most generous with their assistance in finalizing the manuscript and producing the text. Exceptional support provided by Tamara Welschot and Judith Terpos is gratefully acknowledged. Review comments provided by Professor Larry Susskind (Massachusetts Institute of Technology), coeditor of this book series, were most helpful in improving the manuscript and further refining the transdisciplinary and sustainability concepts.

University of Hawaii, Honolulu — Chittaranjan Ray
University of the Pacific, California — Ravi Jain

Contents

Contributors

Rahul Dev c/o Prof. G. N. Tiwari, Centre for Energy Studies, Indian Institute of Technology Delhi, Hauz Khas, New Delhi, 110016, India
e-mail: rahuldsurya@gmail.com, rahuldsurya@yahoo.com

Thomas Grischek Department of Civil Engineering & Architecture, Division of Water Sciences, University of Applied Sciences Dresden, Friedrich-List-Platz 1, 01069 Dresden, Germany
e-mail: grischek@htw-dresden.de

Ravi Jain School of Engineering and Computer Science, University of the Pacific, 3601 Pacific Avenue, Stockton, CA 95211, USA
e-mail: rjain@pacific.edu

Jay Jasperse Sonoma County Water Agency, Engineering and Resource Planning Division, 404 Aviation Boulevard, Santa Rosa, CA 95403, USA
e-mail: Jay.Jasperse@scwa.ca.gov

Ed Pejack School of Engineering and Computer Science, University of the Pacific, 3601 Pacific Avenue, Stockton, CA 95211, USA
e-mail: epejack@pacific.edu

Kishore Rajagopalan Illinois Sustainable Technology Center, 1 Hazelwood Drive, Champaign, IL 61820, USA
e-mail: kishore@istc.illinois.edu

Chittaranjan Ray Civil & Environmental Engineering, University of Hawaii at Manoa, 2540 Dole Street, Holmes Hall 383, Honolulu, HI 96822, USA
e-mail: cray@hawaii.edu

Gopal Nath Tiwari Centre for Energy Studies, Indian Institute of Technology Delhi, Hauz Khas, New Delhi, 110016, India
e-mail: gntiwari@ces.iitd.ac.in

Chapter 1
Introduction

Chittaranjan Ray and Ravi Jain

Abstract The importance of continuing the development of a worldwide clean water supply cannot be overstressed. Developing systems that allow over 6 billion people to have access to 1% of the world's total water volume is no small feat. This is particularly exacerbated by the overuse and misuse of water, and the disparity between affluent regions and those experiencing poverty. While efforts from international organizations during the last fifteen years have provided 1.1 billion people access to clean water which they would not have had otherwise, there remains yet another 1.1 billion people who do not have access to safe water supply (UNICEF Handbook on Water Quality 2008). According to the 2008 UNICEF "Handbook on Water Quality," insufficient water supplies coupled with poor sanitation causes 3.4 million deaths per year (p. 1), which translates into someone dying every 10 sec. Clearly this is a challenging task.

Keywords Contamination • Drinking water • Filtration • Health risks • Pathogens

1.1 Nature and Extent of the Problem

To further grasp the disparity and the magnitude of the water crisis it must first be understood how much water is used by major industrialized countries. For the sake of making a point, the United States will be used as an example. The consumption of the US economy equates to 1,400 gallons per person per day. This includes water used in agriculture, thermoelectric generation, industry, and household use (Hutson et al. 2000). In 1990, the average US resident used between 185 and 200 gallons per day for household use.

Compare this information with the 1.1 billion people who do not have access to clean water: UNICEF estimates that they use 1.3 gallons per person per day (UNICEF

C. Ray (✉)
Civil & Environmental Engineering, University of Hawaii at Manoa,
2540 Dole Street, Holmes Hall 383, Honolulu, HI 96822, USA
e-mail: cray@hawaii.edu

C. Ray, R. Jain (eds.), *Drinking Water Treatment*, Strategies for Sustainability,
DOI 10.1007/978-94-007-1104-4_1,

Handbook on Water Quality 2008). For the world to be on a level playing field, this represents a 99.3% reduction in water use on the part of the United States or an increase in water use of 141 times for developing countries. On a basis of volume comparison, multiplying the average daily use of 304 million Americans (July 2008) and the use of 1.1 billion water-stressed people, the result is as follows: The US uses 55.6 billion gallons per day when the world's water-stressed people use only 1.4 billion gallons per day. That means that 5% of the world's affluent population uses 39 times as much water as 16% of the world's population.

Provision of clean water at an affordable price is also inextricably tied to efforts to erase gender inequality, alleviate poverty, enhance productivity, and afford educational opportunities. Providing access to clean water at an affordable price requires that four major requirements be met: the existence of a source of sufficient quantity, adequate water quality for the intended purpose of use or the ability to increase water quality to meet requirements, a transmission network to a location proximal to usage clusters, and a pricing structure which reflects economic and social capacity.

While it is indeed a "blue planet," fresh water required to sustain and enrich human life is barely 2.5% of the available water (Postel et al. 1996). Two thirds of this water is tied up in glaciers and permanent snow cover leaving a scant 1% to supply the growing needs around the globe. Demands of maturing economies, increasing population, industrialization, and increasing standards of living in many regions of the world are all contributing factors to the current environment of water stress and shortage. At the same time, global fluctuations in climate and a growing imbalance of population distribution between rural and urban centers are adding to the logistical complexity of providing access to water where needed most. Sustainable watershed development, rainwater harvesting, and responsible use of groundwater sources are needed to make access to clean affordable water a reality.

The competing demands for water range from ecological services, food and feed production, power generation, shipping, as well as domestic and industrial needs. While the framework to satisfy each of these demands in a socially acceptable manner can be highly complex and location specific, water conservation as a basic tenet in any such framework is largely non-controversial and a keystone component. Examples of such conservation measures are a more sustainable life-cycle, less water intensive food and industrial production systems and processes, and a more efficient transmission network.

While all of the above mentioned measures—watershed development, production, and protection; judicious water harvesting; and water conservation—will be the predominant tools to meeting the Millennium Development Goals, they need to be supplemented by additional measures to identify, develop, and upgrade alternative sources of water to meet the anticipated gaps in the demand-supply gap in the long run. This is important if water is not to become the critical bottleneck in the development of large parts of the world facing increasing population, dwindling supplies of fresh water, and increased pollution of existing water supplies.

The list of courageous individuals and groups who are addressing, in their small corner of the world, water crisis continues to grow, and those individuals deserve proper credit and acknowledgement. To make the epic proportions of this issue more manageable, this book focuses on several widely accepted water treatment technologies. The development and application of each technology is discussed as well as the various water contaminants eradicated by each technology and the cost of implementation.

1.2 Water Contaminants

It is estimated that 2.6 billion people lack improved sanitation defined as access to public sewer systems, septic systems, pour-flush latrines, or pit/ventilated pit latrines (UNESCO 2009). The interplay between inadequate sanitation and insufficient water is both inextricable and complex, with each exacerbating the other. It's important to keep in mind that water scarcity is often a problem of water quality as well as quantity (Bauer 2004). Water quality is really an issue of sanitation that arises from the widespread presence of contaminants in our waterways. There are many sources of water pollution and most are due to anthropological activities. A few principal sources of contamination are from:

- Discharge of untreated sewage containing chemical wastes, nutrients, and suspended matter. Discharge includes direct input from animals or open sewage sources as well as leakage or poor management of sewage systems.
- Industrial discharge of chemical wastes and byproducts
- Surface runoff from fields (agricultural, construction sites, and other highly permeable zones with high human interference) containing pesticides, herbicides, fertilizers, petroleum products, and other modified or fabricated additives.
- Discharge of heated and/or contaminated water used in various industrial processes
- Atmospheric deposition of contaminants

Some of the major contaminants of surface water include:

- *Escherichia coli* (*E. coli*)
- *Giardia*
- *Cryptosporidium*
- Viruses
- Pesticides
- Organic and synthetic compounds
- Pharmaceutical compounds

Major contaminants of groundwater include most of the above and heavy metals and metalloids such as arsenic and other ions. Larger pathogens such as *Cryptosporidium* and *Giardia* may be filtered out during passage through soil.

Escherichia coli is one of the most threatening pathogenic contaminants in the world, due to its prevalence in water systems unprotected from fecal contamination. Although there are many harmful microorganisms and bacteria that enter our water, most are introduced through animal feces or human sewage and are indicative of total coliforms. *E. coli* is not an exception, and the best preventative action is also one of the simplest: protect water sources from fecal influence. Fecal matter present in drinking water can most often cause gastrointestinal diseases, but may lead to more life threatening diseases.

Giardia, another fecal-released contaminant, acts as a parasite to infect the small intestine and cause diarrhea. It can be found in the stool of infected individuals and can exist in one of two forms: an active trophozoite or an inactive cyst. Cysts have a protective layering and consequently can survive in water systems like fresh water lakes and streams for extended periods of time. Up to 20% of the world's population is chronically infected with *Giardia lamblia* (Marks 2009).

Similarly, *Cryptosporidium* is a parasite that also thrives in the small intestines of calves, others animals including humans, and is released into our water systems. Symptoms include fever, diarrhea, nausea, and cramping. *Cryptosporidium* can cause severe, persistent problems and in some cases, even catalyze death for individuals with highly compromised immune systems from AIDS/HIV or transplant procedures. It is most commonly found in lakes, rivers, and predominately affect those who come in contact with water contaminated by feces (NSF International 2009).

Human enteric *viruses* in drinking water are of concern as they are very small (less than 100 nm) in size and can survive in the environment. Although most viruses are host specific (e.g., a human enteric virus typically attacks humans, not other animals), recent concerns about non-enteric animal viruses affecting humans cannot be understated. The maximum allowable infection risk for humans is 1 in 10,000 persons per year (Regli et al. 1991). For viruses, the dose-response relationship is based on rotavirus and poliovirus 3. The maximum allowable concentration is then 18 viruses in 100 million l of water. Schijven et al. (1996) states that the virus concentration in surface water must be reduced by 5–8 logs to meet the standards.

Pharmaceuticals and personal care products (PPCPs) present in surface water are of concern due to their endocrine disrupting behavior. The primary mode of entry of PPCPs to surface water is from wastewaters. In general, PPCPs are found to be fairly ubiquitous in wastewater, relatively resistant to removal in conventional wastewater treatment plants, and quite persistent in the environment (Daughton and Ternes 1999). In a nationwide study in the United States, Kolpin et al. (2002) presented a detailed picture of the pharmaceuticals found in surface water sources as part of the US Geological Survey (USGS) National Water Quality Assessment (NWQA) program. The "Emerging Contaminants" project of the USGS Toxic Substances Hydrology Program focuses on analytical methods, environmental occurrence, transport and fate, and ecological effects (http://toxics.usgs.gov/regional/emc/). Evidence of endocrine disruption of male fathead minnows due to exposure to wastewater in aquarium studies has been presented by the USGS (Barber et al. 2007). More and more of these PPCPs are being found in rivers in pristine, urban, as well as agricultural landscapes. More recently, the USGS (Phillips et al. 2010a, b) pointed out that pharmaceutical formulation facilities are sources of opioids and other pharmaceuticals to wastewater treatment plant effluents.

Pesticides (a generic term used for herbicides, insecticides, and fungicides) are another group of chemicals that can be found in surface waters in spring and early summer months in rivers traversing agricultural watersheds (Ray et al. 2002c). While the concentrations of atrazine, one of the most common corn herbicides used in the United States can reach 10–25 µg/l in large rivers (e.g., Illinois River or Platte River), the concentration of atrazine can be as high as 50 µg/l in smaller rivers or creeks following rainfall. The current maximum contaminant level (MCL) for atrazine is 3 µg/l. When all other pesticides are added together, total concentrations in large rivers can be significantly high. Although developing countries use much less pesticides than western countries, particularly the United States, land management practices and use of new hybrid varieties are still adding significant pesticide loads to surface waters.

Synthetic organic chemicals are typically discharged by industries to the sewer networks. Typical household contribution of synthetic organic chemicals is typically lower than industrial discharges (mostly from household cleaning solvents). Typical synthetic organic chemicals could be volatiles or semi-volatiles in the US Environmental Protection Agency (USEPA) primary pollutant list or polyaromatic hydrocarbons.

The quality of groundwater is of significant concern for where the groundwater is for drinking purposes. This includes metalloids such as arsenic, pesticides and organic chemicals, pathogens such as bacteria and viruses, and heavy metals. In the above paragraphs, we have addressed issues dealing with synthetic organic chemicals, pesticides, and pathogens.

Arsenic is naturally occurring and therefore can be found in most contexts. It is an element on the periodic table and is most commonly found in four oxidation states: $+5, +3, 0$, and -3. The most threatening and common in water sources are arsenate, As(V), and arsenite, As(III) (Hanson and Bates 1999). Arsenic is introduced into the human body via food, air, and water. Arsenic levels in the atmosphere are considered negligible is most cases, except where power plants are proximal and known to be polluting the local atmosphere with arsenic (Hanson and Bates 1999). It has been reported that on average 10 µg/day of arsenic is ingested through dietary consumption, though this level is not considered toxic (Pontius 1994). Therefore, the highest risk of arsenic poisoning comes from water.

Arsenic exists in most water sources, though it is particularly an issue in groundwater sources. Regions that use pesticides containing arsenic for agricultural reasons are more likely to encounter health issues related to arsenic. Additionally, there are certain parts of the world that have higher levels of arsenic for various geologic and geographic reasons. Bangladesh is one such region where it is estimated that 65% of their 2.5 million water wells have arsenic contamination that exceeds the national limit (Mamtaz and Bache 2001; Munir et al. 2001). This geographic phenomenon is due to the leaching of arsenic from underground rock formations. If the aquifer layer has naturally occurring arsenic then the arsenic can be released into the aquifer during high-draw, low-level periods. Therefore, the geologic impact on arsenic levels is significant as it is a factor of the containing strata of regional aquifers.

Arsenic is also a by-product of copper, lead, and zinc mining and the United States produces approximately 2.5 million pounds of arsenic through smelting processes (NRDC 2009). Additionally, about 90% of the US use of arsenic is found in wood preservatives, and therefore people living in homes with large amounts of preserved wood or near a smelting plant are also more at risk of arsenicosis (arsenic poisoning).

Health risks are sometimes difficult to identify, though there are potentially deadly results. Arsenic is both tasteless and odorless, and therefore it cannot be detected without scientific means (USEPA 2007). For this reason arsenic was a common poison during medieval times. There are both short and long term health effects related to arsenic, though it has been found that a low level of arsenic exposure is essential, and has been included in the USEPA limits (Hanson and Bates 1999).

Short term health threats from high exposure to inorganic arsenics, As(III) and As(V), range from insignificant to lethal. Stomach aches and skin irritations are the result of short-term high-exposure and long-term low-exposure to inorganic arsenic. However, exposure over 100 mg of arsenic has resulted in miscarriage and infertility in women as well as heart disruptions, brain damage, nervous system damage, and DNA damage in both men and women. Arsenic is the twentieth most abundant element on Earth and the most toxic (Lenntech 2009).

Arsenicosis is the most common category of health effect from arsenic exposure, and is a result of long-term, low-exposure consumption, usually between 30 and 800 μg/day. The effects of arsenicosis include decreased white and red blood cell production, lung irritation (and possibly lung cancer), skin irritation and welts, and cancer to various organs (Hanson and Bates 1999; Lenntech 2009). Unfortunately, for many regions of the world that do not have access to scientific water testing, many users begin demonstrating the effects of arsenicosis before arsenic is discovered to exist in their water source. Historically, regions would not know arsenic existed until a fair population had demonstrated illnesses related to arsenic poisoning.

There have been many systems and technologies developed to remove arsenic from water. Mentioned here are only technologies which are relatively low-cost and easily deployable on a small scale:

- Activated alumina filtration
- Complex iron matrix (CIM)
- Manganese greensand filtration (MGF)

Arsenic is most toxic in a compound form, usually $H_2AsO_4^-$ and $HAsO_4^{2-}$, and therefore requires the oxidation of a metal to release the arsenic from the hydrogen/oxygen bonding and immobilization on the metallic oxidizer. It has been demonstrated by Hussam and Munir (2007) that reactions with cast iron provide significant immobilization of arsenic. This process was demonstrated in the SONO filter, now used by approximately half a million people in Bangladesh. During their study, Hussam and Munir began with initial concentrations of arsenic between 32 and 2,423 μg/l with results between <2 and 8 μg/l, well below both the USEPA and WHO standards. The SONO filter is a two-bucket system that utilizes a CIM, sand, charcoal, and brick shards. This combination not only removes arsenic by immobilization within the cast iron complexation, but it also removes coliform from the water without the need for backwashing, cleaning, or leaving residual traces of contaminants. Additionally, the longest SONO filter has been in use for over five years with a total production of 125,000 l with no arsenic breakthrough. Ideal flow rate is between 20 and 30 l/h, significantly higher than other technologies discussed which remove other forms of contaminants, though design tests demonstrate that the SONO filter can operate effectively up to 60 l/h.

Activated alumina filtration was shown by Deb et al. (1997) to be effective in removing arsenic to less than 50 μg/l from initial concentrations of 100–250 μg/l. This method is only effective with an initial pH of <8.2.

It was also found by Hanson and Bates (1999) that a method called MGF was also effective in removing arsenic to levels less than 4.2 μg/l with a filtration time of 15 min–6 h. This method utilizes a complicated chemistry to remove arsenic through the activation and deactivation of glauconite and potassium permanganate. Therefore, the greatest disadvantage presented in this removal method is the high level of technical competency required to construct and maintain a filter.

Lead is another one of Earth's natural metals that contaminates our water systems as a result of human activities. It is found in many metal products including batteries, ammunition, gasoline, paint, and ceramic products. However, it is never found in water naturally and most often times gets there as a result of the corrosion of brass and copper delivery systems. "It is estimated that lead in drinking water contributes between 10 and 20% of total lead exposure in young children" (NSF International "Fact Sheet: *Cryptosporidium* and Drinking Water from Private Wells" 2009). Lead exposure can cause memory problems, anemia, and in severe cases lung, brain, and kidney damage.

Minor amounts of *nitrate* and *phosphates* can occur naturally in surface water systems, but harmful levels are introduced by improper disposal of human waste, fertilizers, septic systems, animal feedlots, industrial waste, etc. When nitrogen combines with oxygen or the ozone, nitrates are formed. High levels of nitrate-nitrogen in drinking water can be very dangerous to human health, especially for pregnant women and children. In addition, nitrates are made in excess amount by plants, animals, automotive, industrial, and smoke exhaust.

"Per capita we contribute approximately 3.5 pounds of phosphate yearly to our environment" (Rail 1989). Nitrate exposure leads to a blood disorder known as methemoglobinemia symptomatic of vomiting, diarrhea, and "blue baby" syndrome, a breathing problem in children under the age of five. Excessive phosphate exposure can cause kidney problems and osteoporosis.

While this is not a major problem for rural communities in developing countries there are enormous amounts of methyl tert-butyl ether (MTBE) and pharmaceuticals entering water systems in some industrialized states in the United States; some of these act as "endocrine disrupters." These disrupters have been correlated with developmental, reproductive, and other health problems in wildlife and laboratory animals. The two main sources for *pharmaceuticals* entering our waterways are from homes and hospitals. They most often enter the water through incorrect disposal of partially used or expired medications and partially metabolized excretion of medications through human waste. While wastewater treatment plants are designed to remove contaminates from influent water, many treatment systems cannot remove pharmaceutical contaminants.

Described above are major drinking water related contaminants. Their serious health implications and the extent to which populations are exposed are discussed. Suggested drinking water treatment processes and their analysis discussed in this book provide options for implementing appropriate and sustainable technologies for water treatment. Clearly, the protection of water resources from contamination—human, agricultural, and industrial—is an essential step to providing affordable and safe drinking water.

1.3 Topics Covered

Technology continues to be a primary focal point in the effort to provide clean water to the world's population. The complexity of developing technology, which can support itself while bridging the technology gap between developed and developing countries, remains one of the greatest challenges. Coupled with the social and intellectual challenge of knowledge transfer, the capacity for technology to solve systematic problems becomes entangled in a web of questions whose answers have been the pursuit of scientists and engineers for decades. Because of these questions, this book also discusses some of the key factors that make a technology desirable and sustainable in developing countries.

Several technologies have surfaced as reliable sources in many regions of the world. This book focuses on only four areas of water treatment: natural filtration, membrane filtration, solar distillation, and solar pasteurization. Each represents a different scale of application and appropriateness. Within each area exists a multitude of variance and diversity as creative minds continue to develop modifications to better address context-specific issues.

Chapter 2
Drinking Water Treatment Technology—Comparative Analysis

Chittaranjan Ray and Ravi Jain

Abstract Water treatment technologies have evolved over the past few centuries to protect public health from pathogens and chemicals. As more than a billion people on this earth have no access to potable water that is free of pathogens, technologies that are cost effective and suitable for developing countries must be considered. Sustainable operation of these treatment processes taking into consideration locally available materials and ease of maintenance need to be considered. In this chapter, we consider natural filtration for communities of various sizes. In natural filtration, slow-sand filtration and riverbank filtration are considered. Slow-sand filtration is suitable for small to medium size communities, whereas riverbank filtration can be suitable for small to very large communities depending on site and river conditions. Membrane filtration is another technology that can have application to individual households to moderately large communities. Both pressurized and gravity-fed systems are considered. For the developing regions of the world, small membrane systems have most applications. Solar distillation is a low-cost technology for sunny regions of the world. Particularly, it has the most application in tropical and semi-tropical desert regions. It can use low quality brackish water or groundwater for producing potable water. These systems can solely operate with solar energy. The scale of application is for individual households to very small communities. Solar pasteurization, like solar distillation depends on solar energy for purifying small quantities of water for individual or family use. It is most suitable for remote, sunny, high mountain regions such as the Andean mountains, central Africa or the Upper Himalayas where electricity is not available. Also, reliance on firewood is not feasible due to barren landscape in many of these regions. Also, case studies of natural (riverbank and lakebank) filtration, membrane filtration, solar distillation, and solar pasteurization are presented.

Keywords Natural filtration • Solar distillation • UV radiation

C. Ray (✉)
Civil & Environmental Engineering, University of Hawaii at Manoa,
2540 Dole Street, Holmes Hall 383, Honolulu, HI 96822, USA
e-mail: cray@hawaii.edu

C. Ray, R. Jain (eds.), *Drinking Water Treatment,* Strategies for Sustainability,
DOI 10.1007/978-94-007-1104-4_2,

2.1 Introduction

The goal of all water treatment technologies is to remove turbidity as well as chemical and pathogenic contaminants from water sources in the most affordable and expedient manner possible. Many technologies, which have been developed, work best in demand-specific contexts: either the demand of mass-volume or of mass-flow. In all technologies discussed throughout this book, the sun's energy or the soil's filtration capacity or energy efficient membrane filtration are the primary mechanisms of purification. The main components which will be compared in this section include flow rate (m^3/day), cost of implementation, maintainability (which includes cost of maintenance, availability of spare part and materials, and technical knowledge required for repairs), energy consumed (either MJ/h or kW/h), and reliability (as a function of total number of serial components and the sensitivity of each component to long exposure to adverse conditions).

While discussing technology, it will be important to keep in mind the ethic of engineering water systems, acknowledging the social re-shaping which occurs inherently within design implementation. As stated by Priscoli (1998) the answers to water management systems "depend, to a great degree, on what you want or think the ecology ought to be." (Priscoli 1998) He outlines four main views of technological intervention: gigantism, technological triumphalism, historical romanticism, and techno-phobias. The first two reflect mindsets, which hold technology in too high a regard, with gigantism referring to massive infrastructure installation and triumphalism referring to some enigmatic future point where technology becomes superior to nature. The latter two views debase the value of technology and its ability to address water issues around the globe with romanticism quoting partially-factual events and systems in the past and criticizing present uses of technology, and with phobias technology is never a correct answer as it replaces the "natural way" to some degree. Therefore, as each technology is discussed and mentioned throughout this book and as users consider implementation of specific technologies, it will remain important to be aware of ecological and ethical impacts of the technology. The spatial applicability of the technologies varies widely. While some are more appropriate for communities (cities or towns, e.g., natural filtration), others are more appropriate for families or individuals (solar pasteurization, solar distillation).

2.2 Natural Filtration

Perhaps the most ubiquitous of treatment technologies humanity has employed is natural filtration since the beginning of written history. In Exodus 7.24 of the Bible states "And all the Egyptians dug round about the Nile for water to drink, for they could not drink of the water of the Nile." Thus implying the hole on the bank provided clean water relative to the contaminated water of the Nile. Quite simply, natural filtration takes advantage of the soils that act as filters as the water passes through them. It is important to note that there is a difference between drawing groundwater from aquifers and utilizing natural filtration to produce drinking water when the water comes from a surface source. The groundwater in aquifers results from a form of natural filtration of rainwater. There is also riverbank filtration, which is drawing infiltrated river water as it migrates toward pumping wells in adjoining alluvia aquifers, and there is constructed slow-sand filters, which take advantage of the natural filtration attributes of well sorted soils in a constructed—and therefore well contained—environment.

The most common form of natural filtration used currently is sand filtration in a natural setting. However, as more studies are being published on the advantages of true natural filtration, more projects are being undertaken to utilize riverbank filtration. Also, simple wells can be classified as using natural filtration, assuming the soil isn't contaminated and most of the water drawn from the well is a result of rainfall infiltration, but discovering answers to those questions are more technically demanding than this introduction section, and will therefore not be discussed. The focus shall remain on water treatment technologies and not water supply technologies because unless wells are located within a reasonable distance from an open water source, they are too ambiguous as to source and purification attributes.

The best materials to be used for natural filtration are unconsolidated alluvial deposits due to high hydraulic conductivity. The greatest disadvantage of using unconsolidated soil is that there is the possibility of the introduction of anthropogenic contaminants from the land surface to groundwater (typically alluvial aquifers are unconfined aquifers). However, there are clear advantages: natural filtration of appropriate travel time can induce a 3–5 log reduction in microbes and protozoa (Schijven et al. 2002). A 1 log reduction represents a 90% removal of the bacteria or protozoa. Therefore, a 3–5 log reduction removes all unwanted biological and viral components from water to an undetectable—or at the very least, an acceptable—level. However, due to the changing redox conditions, there are often increased amounts of manganese and iron in naturally filtered water, as well as the formation of some sulfurous compounds that are malodorous. These negative effects are eliminated when using rapid sand filtration, but the advantages are also subdued, as will be seen in the section below on sand filtration (Hiscock and Grischek 2002).

2.3 Riverbank Filtration

Surface water in river systems is dynamic: it is flowing downstream, it is evaporating or taken up by riparian vegetation, it is infiltrating into groundwater (or it is entering the river from the groundwater through its bank and bed), and its ability to do all of this is highly impacted by the geologic composition of the immediate environment. There is also a dynamic interaction between surface and groundwaters in natural settings. When the river floods, water from the river gets stored in the soils in the bank areas and the low-lying areas between the floodplains. When the river level drops, the stored water from the bank areas slowly drains back to the river.

Riverbank filtration takes advantage of the infiltration of river water into a well through the riverbed and underlying aquifer material. This is a natural filtration process in which physico-chemical and biological processes play a role in improving the quality of percolating water. After a certain zone of mixing and reducing, the infiltrated water is at its cleanest: almost all river contaminants are removed. Wells are installed in this zone to pump the water to be used for drinking. The purity of this water and its suitability for drinking is outstanding, even in examples where there is an event that introduces a shock load of contamination into the river. Due to the geologic media's ability to remove the contaminants and travel time of water abstracted for natural filtration, the impact of such an event is minimal and requires minimal treatment to address.

The size of riverbank filtration systems vary widely—some systems producing less than 1 million gallons per day to others producing hundreds of millions of gallons per day. The production at a site depends on the utility's need, number of wells at the site, type of wells and pumping capacity of each well, local geohydrology, hydraulic connection between the river and the aquifer, distance and placement of the wells from the river, and a host of other factors. Ray et al. (2002a, b) provides comparative production of water at various RBF sites.

In a natural environment, the variations in production from RBF wells are caused by two main factors: local hydrogeology and river hydrology. While it is critical to consider the hydraulic conductivity of the aquifer, one must also consider the river hydraulics such as grain size of the clogging layer, shear force against the riverbed (to gauge erosion, transportation, and deposition factors for clogging), mean velocity, the hydraulic gradient line, and flood peaks. In addition to local hydrogeology and river hydrology, it is also important to understand catchment zones and other sources of infiltration in the broader geological region affecting the site. The result of these factors combined is a rather tedious and technical scenario, which requires immense amounts of research before being able to confidently draw pure water.

However, riverbank filtration has been used for 130 years in Germany (Schubert 2002a, b) yet it wasn't until the 1980s that any significant amount of research was published beyond the water utilities operating the RBF systems in regards to the parameters mentioned above. Therefore, despite the technical complexities of developing a well-understood site, there are general and basic parameters that can be very simply employed to ensure water quality from riverbank filtration. Three very easily identified parameters are river condition, soil and aquifer composition, and well location.

Due to riverbed clogging (often termed colmation in Europe), it is best to develop riverbank filtration sites in areas where sediment transport is taking place. Also, regions that are experiencing erosion tend to not have as deep alluvial materials to extract the water from, again making regions of sediment transport preferable for developing riverbank filtration systems. This region is common in foothills and valleys and is generally characterized by large bends in the river, and low to moderate flow velocity (0.5–2.5 m/s) depending on sediment load and riverbed composition. As stated previously, the best conditions for riverbank filtration are in unconsolidated alluvial deposits (although there are examples of low-permeability zones being used for natural filtration (Ray et al. 2002b; Hubbs 2006). Wells used in the alluvial aquifers have used a variety of technologies for installation. For example, vertical wells or horizontal collector wells used in western countries use mechanical means for drilling and installing laterals (screens). While modern technologies are being used in India currently for the installation of new vertical or horizontal collector wells, many operating large collector wells (e.g., those at Hardwar, India) were built manually by digging the soil and making the caisson and installing the gravel and cobble pack around the port openings. Most Indian companies still use direct push technology for installing laterals in which the screen pipes with open holes are pushed directly into the aquifer.

Groundwater pumped very near a surface water source may contain contaminants found within the open water source. However, groundwater pumped at a long distance from an open water source can be affected by contaminants that are typically present in groundwater. Therefore, there is an identified "mixing zone" associated with each surface water source. This zone is defined as the zone where contaminants from surface water have been removed without the addition of groundwater contaminants. Zone width is a function of hydraulic conductivity, and is dependent on mean travel time, with targets between 5 and 20 days. Therefore, it can be reasonably attributed that on the scale of technologies being compared in this book, riverbank filtration does not demand excessive technological expertise to develop or maintain, particularly in regions without any access to any form of filtered or cleaned water source.

Use of multiple wells and redundancy are some of the common ways to ensure steady supply of water during repair and maintenance of wells or during mechanical failures of pumps or well rehabilitation. Multiple wells constitute a parallel process where one or more wells can be off line and the system can still meet demand. However, simply because there are so many individual wells involved introduces the chance of failure and therefore maintainability becomes a larger issue, particularly for regions without access to surplus manpower or materials for repair. When multiple vertical wells are used in riverbank filtration systems, the pumping efficiency can be increased by installing a siphon system and pumping the water from the caisson where the siphons empty the water from multiple vertical wells. Such a system is operated at Düsseldorf, Germany.

Due to the use of large mechanical pumps, riverbank filtration relies on either an electrical power grid or internal combustion engines to provide enough energy to the system for operation. There is also a dependency on larger infrastructure as many sites utilize multiple wells, and must therefore be connected to a common storage point or multiple storage points. Either way, the system-wide maintenance demand is larger than what is required for slow sand filtration (another natural filtration system), but less than the requirements for membrane filtration based on the size of the compared systems. Since the only distillation and pasteurization discussed in this book are solar-powered technologies, it is difficult to compare the energy consumption of a system that could be solar but may also very likely be diesel or electrical-grid. However, even if PV panels were used to supplement energy needs, the area of panels required to power well pumps can sometimes exceed practicality depending on well depth, hydraulic conductivity, and topographic/weather allowances for PV arrays, along with human/livestock complications of installing a relatively large solar array. Therefore, it is not unreasonable to assume that riverbank filtration will depend on a pre-existing electrical grid or diesel/biofuel/hybrid generators. Since this is the only technology that inescapably requires pumping (while some others may need it only in certain conditions), riverbank filtration requires more absolute energy than most other technologies considered. This, however, does not include delivery systems, which may often require additional pumps, or appropriately scaled pumps to handle both withdrawal and distribution. Therefore, the total combined energy required for any system will also be a function of the service area of that system.

Use of large pumps is one of the key considerations in riverbank filtration systems. Newer systems use variable drive pumps that require cool (air conditioned) environments to operate. Other technologies, with the exception of various membrane filtration technologies such as reverse osmosis (RO) do not have need for pumps. Sand filtration or solar distillation has low energy. Also, unlike solar energy dependent technologies, riverbank filtration would have dependable supplies due to the use of electric or diesel motors.

Conversely, compared to other forms of water treatment technologies for large systems, riverbank filtration is one of the easiest to implement due to relatively low technology demands and simplicity of construction, training, and operation. In this, it is meant that the concept of drawing water from the ground is as old as history, and therefore justifying digging wells near a river is quite easily done. Convincing locals that water from the well is more pure than river water may require some work, since the work of purification by soil is not easily observed by users. Riverbank filtration also has the capacity to begin at a smaller scale to demonstrate the purity of water drawn and later expanded into a larger scale due to its parallel nature. In fact, many utilities operate a pilot well a year before building a full-scale system.

Manpower to dig wells is available around the globe. Pumps of various levels of technology can be found in almost as many places as Coca-Cola®, and the training required to understand how to use a pump/well system is almost minimal due to their pervasive use. This allows a technology to be introduced that minimally alters expectations, can be easily understood, can be scalable, and can have tangible, observable results. The combination of these attributes makes riverbank filtration an attractive option to introduce to regions with access to contaminated surface water, but little or no access to purified water.

Potentially one of the challenges facing riverbank filtration is water-rights mitigation and legal intricacies. This only affects regions with water-rights policies (e.g., western United States), but increasingly more of the world is affected. However, many riverbank filtration systems are successfully operating in the Western United States. Therefore, it is important to consider the potential legal ramifications of implementing a system that removes water from a broader, underground source.

2.4 Slow Sand Filtration

Slow sand filtration is a fabricated form of natural filtration, which is created within a man-made context for the specific purpose of filtering water. This filtration method has been municipally used since the nineteenth century, and continues to be an excellent filtration method. As stated by the World Health Organization's Water Sanitation and Health (WHO WASH) division in their 1974 report *Slow Sand Filtration*, "Under suitable circumstances, slow sand filtration may be not only the cheapest and simplest but also the most efficient method of water treatment" (Huisman and Wood 1974).

Constructed from simple materials such as wood or even a modified shipping container, slow sand filters are basic enough to be adaptable to a wide range of available materials. The filter itself is usually 1 m thick, with a minimum of 0.7 m of fine sand. The remaining portion that isn't sand is gravel and pebbles located at the bottom of the filter to allow the purified water to collect and drain from the container. The filter is then filled with water until saturated, and there must also be supernatant water on top of the sand in order to cultivate and sustain the *Schmutzdecke*. There are no mechanical components, and no electricity is required to operate. Gravity is the external force, and the natural bacteria and protozoa within the *Schmutzdecke* actively treats the water.

It is, in fact, the *Schmutzdecke* that is responsible for nearly all the filtration that happens. Quite literally, it means "grime or filth" in German, as it is a small biofilm which forms at the sand-supernatant boundary consisting of naturally occurring bacteria and other organic compounds, which interact with the water as it passes through. It is this interaction that is able to filter out particles smaller than the inter-granular space created by the sand and other biodegradable contaminants; and therefore it is much more efficient at purifying water than rapid sand filtration.

Rapid sand filtration is simply a slow sand filter without the *Schmutzdecke* (or biofilm) and is typically employed at a majority of water filtration plants. Therefore, the only filtration that occurs is due to the sand particles hindering large suspended colloidals from passing through the intra-granular space and to some physico-chemical interactions between the sand and the contaminants. It cannot purify water nearly to the degree slow sand filtration and riverbank filtration can, and for its efficiency it requires frequent backwashing. Backwashing is an engineering challenge for systems that operate on low technology. Often other processes such as coagulation, flocculation, and sedimentation are employed before engineered filtration using rapid sand filters. Thus, it is not considered a "natural" filtration system.

Cleaning of slow sand filters takes place between once every three weeks and once a year, depending on the quality of the raw water source. It is also well within grasp of an ordinary citizen, though knowledge of how the process is actually cleaning the filter is helpful. Additionally, in order for a slow sand filter to be fully operational, it requires 1–2 days for the biofilm to form, and until then the filtered water is not usually suitable for drinking, and must therefore be recycled through the filter until a full biofilm is in place. Then, the biofilm continues to grow throughout the use of the filter and must therefore occasionally be cleaned. When the biofilm becomes too thick, it begins to impede the flow rate of the filter, and when head loss has reached the design flow maximum, the filter must be cleaned. While there are several methods used to clean slow sand filters, only one will be mentioned now. Referred to as "mechanical scraping" the name can be misleading unless the scraper is automated. In remote locations, the filter can be cleaned by draining the water, drying the sand, and scraping the *Schmutzdecke* off the top layer via manual labor. Then, with a fresh surface of sand for a new biofilm to grow upon, water can be re-applied with the necessary 1–2 days growth period required.

The drawback of sand filters is their inability to fully treat highly turbid water. It is quoted that water with a turbidity of 10 NTU or higher cannot be adequately treated by sand filtration, and water with turbidity enhances the life of the filter and reduces clogging (Tech Brief: Slow Sand Filtration 2000). In order to reduce the turbidity of water, settling tanks may be utilized or even developing several pre-filtration sand-sieves to remove larger particles or aggregates. Due to utilizing the ecological interactions of living bacterium, slow sand filters are not ideal for year-round use in cold climates when the bacteria may become dormant in winter months. In such situations membrane filtration or solar purification may be more appropriate.

An additional drawback lies within the name of this purification method, as it is indeed a slow filter. Flow from a slow sand filter range from 0.015 to 0.15 m^3/m^2h, which can be as much as an order of magnitude lower than other technologies' per unit area output. Thus, for large cities, large filter beds are needed. Storage is also required to mitigate peak demand, and therefore maintaining the purity of the stored water is an introduced maintenance factor.

However, due to their simplicity and size, slow sand filters also have several advantages. Technologically speaking, they are the simplest technology considered, which aids in minimizing maintenance and expediting the education of the community users. Also, due to the fact that the entire system can be very easily self-contained, sand filters are easily scalable. Implementing a sand filter with a surface area of 1 m^2 would not be complicated by expanding to a 10 m^2 basin as long as there is a minimum of 0.7 m of fine sand, and time is given for the biofilm to form.

Material access to sand, gravel, and materials to construct the basin within is widespread, with perhaps the caveat of the drainage plumbing. Additionally, the financial cost associated with the materials, installation, and maintenance is significantly lower than anything else mentioned in this paper. Such an inexpensive project is easily funded by non-profit or microfinance organizations. Additionally, due to the low cost of sand, the installation cost per square meter decreased rapidly with an increase in filter area.

Some studies have concluded that slow sand filtration requires a large footprint (Huisman and Wood 1974). While slow sand filtration is quite practical for large scale applications, it is perhaps even more practical among individual and small community users. Under small-volume demand, the footprint needed for slow sand filtration reduces considerably. While some municipal plants have 200 m^2 of filter area, some home use as small as 10 m^2 (Tech Brief: Slow Sand Filtration 2000). And while riverbank filtration utilizes space underground, and therefore could be argued to use less space, if the region has water-rights laws in place, the size of land area for riverbank filtration becomes quite a serious consideration.

Interestingly, some studies have come to contradictory results. In the Nainital region of northern India (see case study below), it was found that rapid sand filtration was sub-par compared to natural filtration, as it did not remove nearly enough coliform or COD to meet national standards (Dash et al. 2008). However, in a study done by the University of New Hampshire (2003) of five locations in the eastern United States it was found that slow sand filtration was more successful at removing coliform and *E. coli* than natural filtration. It was found, however, that natural filtration was superior in removing dissolved organic compounds (DOC) (41–85% as opposed to 8–20% for sand filtration) and total organic compounds (TOC) (55–75% as opposed to <30%) (Partinoudi et al. 2003). Therefore, it can be generally concluded that when using slow sand filtration (and not fast sand filtration), it is better in an environment that is dominated by protozoa as opposed to a system that has high levels of other organic compounds.

By way of comparing the differences between riverbank filtration and slow sand filtration, the United States Environmental Protection Agency (US EPA) has developed "purification credits" for both technologies as a gauge of how well they meet US EPA drinking water standards. Slow sand filtration is given a log-reduction credit of 3-log removal for *Cryptosporidium* (Ray et al. 2002b). The US EPA requires a 2-log removal for *Cryptosporidium*, and even under the new Long Term 2 Enhanced Surface Water Treatment Rule (which increases the required log reduction by 1–2.5 log), therefore RBF should be used as a "pre-filtration" technology aimed at reducing the load placed on slow sand filters or other filtration devices.

As was stated by the World Health Organization (Huisman and Wood 1974), and reaffirmed through this brief analysis, slow sand filtration is an attractive purification method for situations where low technology and low cost are required, but high quality output is demanded. Its drawbacks are the slow rate of filtration and the inability to purify water with high turbidity; notwithstanding, slow sand filtration is a technology worth considering in virtually any project scenario. We have limited our discussion to riverbank filtration as the sole natural filtration in this book.

2.5 Membrane Filtration

Membrane filtration technology is simply the filtering of water through a sieve or semi-permeable layer such that water molecules are allowed to pass through, but bacteria, chemicals, and viruses are prevented from passing. The sophistication of membrane technology ranges from using a sand-filled T-shirt fed by gravity to highly advanced pressurized systems relying on nano-technology to actively screen microbes.

2.5.1 Pressurized Systems

The most effective membrane technology, pressurized systems, often require significantly more energy than other membrane systems due to electrical or mechanical systems required to maintain the pressure in the system. Yet because of the pressure introduced to the system, the pore spaces in the membrane can be significantly smaller, allowing higher removal rates of contaminants. The most common application of membrane technology is in RO desalination although the application of membrane technology has been used for bacterial and protozoan removal as well. Other desalination processes are membrane filtration (nanofiltration [NF], ultrafiltration [UF], and microfiltration [MF]) and electrodialysis (ED). All three membrane filtration systems are pressurized membrane systems primarily used to purify seawater or brackish water (water containing less salt that seawater, but still more salty than WHO regulations).

Reverse osmosis is used to take saline water and convert it into pure water. It currently makes up 80% of desalination plants for a cumulative 44% of all desalinated water volume (Greenlee et al. 2009). The technical measure of fresh water is to contain less than 1,000 mg/l of salts or total dissolved solids (TDS) and the World Health Organization has established a baseline of 250 mg/l, which is also supported by the US EPA (LT2ESWTR 2006). Therefore, any water containing higher levels of salts or TDS must undergo some sort of removal process.

The energy required for RO is significant due to the nature of the membrane surface (Fig. 2.1). Since RO membranes are considered non-porous, diffusion is the primary transportation function for water to pass from high concentration to low concentration. As stated in Table 4.3 (chapter on desalination) seawater RO requires approximately 3–6.5 kWh/m^3 to reduce average salinity (36,000 mg/l) to below drinking water standards (800 mg/l). While this is significantly less than other desalination technologies discussed in Chap. 4, it is also higher than other technologies compared in this section. As an example, according to Srinivasan (1993), the

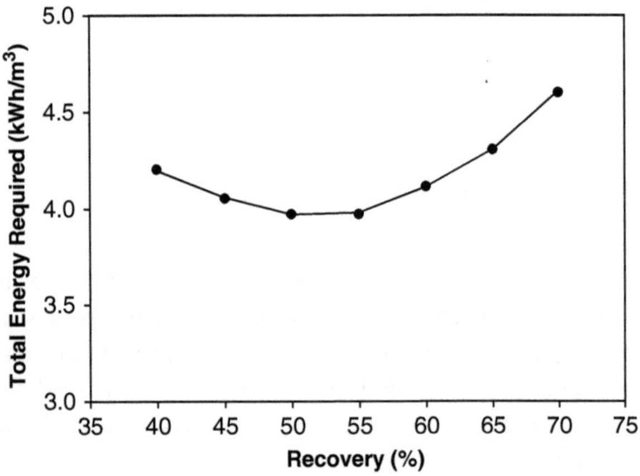

Fig. 2.1 Total energy required per volume of permeate as a function of RO system recovery. (Source: Greenlee et al. 2009)

average solar energy in India is 5 kWh/m^2/day and therefore RO would only be available in certain more sunny regions of India if PV were to provide the energy source.

Since the energy is proportional to the membrane permeability and the feed pressure, the specific kWh/m^3 energy consumption is determined by the feed water make-up, drinking water standards, and flow criteria. Another large component is the recovery rate the RO system is designed to operate under. The trade-off with recovery rate is that a higher rate usually means more saline passage through the membrane, but results in higher outflow.

The cost of membrane desalination is dependent on location (water quality, geography, local economy, etc.) and volume (of required treated water). Greenlee et al. (2009) reported that large RO desalination plants (3,500–320,000 m^3/day) have a cost between $ 0.53 and 1.94/m^3 where as small plants of 0.1–1.0 m^3/day have costs between $ 30 and 36/m^3. However, much of the high cost of small RO plants is due to research instrumentation, which would not be present in on-site installations. Estimates say 40% of the cost will be reduced when implemented in the field.

The pressure required for RO to occur is significant. First, natural osmotic pressure must be overcome by increasing the hydrostatic pressure of the system. For seawater, the osmotic pressure ranges between 2,300 and 3,500 kPa. To overcome this, many RO plants utilize between 6,000 and 8,000 kPa of pressure.

A turbidity of less than 0.2 NTU is recommended for RO systems as fouling occurs at higher rates. The capacity for a membrane to be fouled is exponentially related to the amount of particulates in the feed water. This demonstrates that RO is much more sensitive to particulate than slow sand filtration, requiring higher levels of pretreatment.

The overwhelming majority of technical papers and research articles produced on membrane filtration focus solely on desalination. However, the use of membrane filtration for pretreatment of RO plants is becoming more common. This is no different than simply using the same pretreatment technology to purify water that has no salt concentration for drinking. Pretreatment for RO can utilize various options, but of most interest to this section is the use of MF, UF, and NF. The differentiation between each is the pore size of the membranes (as they are considered porous, unlike RO membranes), with MF being the largest pore-size and NF being the smallest. The ability of each to filter out contaminants is beneficial in various environments, and the correct application of membrane pore-size is largely dependent on the most common contaminants in the feed water.

UF has surfaced as the most common choice for RO pretreatment as it balances the screening capacity of nanofiltration with the flux capacity of MF. NF is primarily used for brackish water and dissolved organic compounds. This is a unique divergence beginning from the classic use of membrane technology, as it represents a more standard water treatment technology as it moves away from strict salt-removal uses.

Another treatment technology used for brackish or salt water is distillation. Distillation can also use a membrane, such as membrane distillation (MD), which utilizes membrane pore-sizes similar to MF, UF, and NF to purify water. The principle of MD is to create a temperature gradient between the feed and permeate sides of the membrane, so water is pulled through on the basis of liquid-vapor equilibrium. This is achieved by using the vapor pressure as the mechanism that moves water from one side to the other. The advantages of MD are that it requires less pressure than RO and less heat than multi-stage flash (MSF) distillation or multi-effect distillation (MED) technologies. Additionally, it can be used in a wider range of applications (such as sustainable water treatment), and can be easily combined with renewable energy sources such as PV panels or wind generated energy (Al-Obaidani et al. 2008).

The cost of MD is quoted to be \$ 1.17/m^3 for a hypothetical plant producing 24,000 m^3/day. However, in conjunction with solar power and a heat recovery system, the price per cubic meter drops to \$ 0.56/m^3, which is similar to that of RO (Al-Obaidani et al. 2008). The same challenge that faces RO would also face MD: as production size is reduced, the cost per cubic meter increases significantly. Karagiannis and Soldatos (2008) report that for plants producing between 2 and 3 m^3/day the cost of seawater desalination is between \$ 3.40 and 6.90/m^3. For communities outside the reach of metropolitan infrastructure and without a sufficient population to justify large plants (or the volume of source water to feed such large plants), it is necessary and practical to create small-scale plants. Additionally, through the development of sustainable components and creative local design, the cost of water would continue to fall. It is good to keep in mind that the volume of permeate water is orders of magnitude less on small systems than on large and as such the aggregate cost is significantly less. For a large volume plant, daily operation costs are between \$ 25,000 and 50,000/day, whereas for small volume plants daily costs are around \$ 25–50/day, significantly less cost for the community, even though the per cubic meter cost is higher.

2.5.2 *Gravity-Fed Systems*

Gravity-fed systems are almost too simple to be worth mentioning, but it is good to be familiar with them since sometimes they are sufficient to purify local water sources. Most often, these systems are used in conjunction with slow sand filtration where the membrane is the medium used to support the sand. Often, large-pore membranes such as cloth fabric or canvas are used. Clearly gravity-fed systems are not designed for high-concentrations of contaminants, but rather to be used as a cheap pre-filter of large suspended colloidal matter in source water.

The cost of gravity-fed systems is so low that it is rarely recorded. Often, the components which make up the system are collected from what the community has on-hand or are purchased at a common convenience store by someone visiting a local city, and are therefore significantly below the scale of costs discussed in this book. The sophistication and corresponding cost of gravity-fed membrane filtration compared to other technologies discussed is similar to attempting to compare a child's lemonade stand with a MinuteMaid® factory. However, it deserves consideration, because if a community discovered that large suspended solids are the primary contaminant, gravity systems may be sufficient to meet their needs. There is often excitement about the implementation of some new technology because of its advanced design and capacity beyond what is expected, but sometimes this excitements leads to overspending beyond what the community has the capacity to support. Therefore, the cheapest options are sometimes the best, even if re-visiting the site several years in the future is a necessary part of implementation.

2.6 Solar Distillation

Before beginning a discussion on water treatment systems that utilize solar power, it is worth mentioning the sun and how much power is actually available. As the Earth is an imperceptible cosmic dot from the sun's perspective, very little of the total energy emitted from the sun ever reaches the Earth. In fact, at the outmost reaches of our atmosphere we receive only one-billionth of the energy that the sun produces. The sun's energy per unit area is called solar flux, and is generally measured in W/m^2. While the extraterrestrial solar flux (flux at the outer edge of our atmosphere) is 1,353 W/m^2, this can never be reached on the Earth's surface. If the solar flux were that high on the Earth's surface we would be in much greater danger from the sun, so we are quite thankful that the atmosphere absorbs much of the solar flux. However, the interference from the atmosphere complicates solar technologies. Due to atmospheric diffusion, solar flux is reduced by at least 15–30%, even on the sunniest day of summer on the equator. Typically, solar flux from 300 to 1,000 W/m^2 is referenced as being used for solar technologies. Often times, references to higher solar flux values include the magnifying characteristics of compounding reflectors.

Solar technology is surprisingly fickle as it is heavily dependent on sufficient solar flux. Attributes that affect solar flux are absorption and scattering by the atmosphere, the time (day, month, or year), latitude, altitude, and meteorological effects. Additionally, technology used to capture the sun's energy is expensive to manufacture and produce, though often not as expensive as other water treatment technology costs.

Under current systems and operations, desalination costs are substantial for developing communities—particularly those with comparatively small populations. The infrastructure required to produce and support continuous desalination and purification—including power supply, pre-treatment, brine management, janitorial maintainability, repairs and modifications maintainability, and inventory—is a daunting task when the protective hedge of other city-sized systems are far removed. However, while cities may have the cash flow to employ full-scale operations to alleviate water needs, those left beyond the reach of urbanization have hand-collected water from unsanitized sources as their only recourse. Yet despite developing countries with 50–70% of their population living in the few urban centers (UN DESA 2007), there still remains hundreds of millions of people qualified by the UN as being "water-stressed" who need access to cheaper and more reliable technology to bring them clean water.

Solar distillation is a rising star among such technologies. A very simple technology in both concept and design, solar distillation utilizes the natural process of evaporation to capture purified water. The structure used in solar distillation is called a solar still, and a common solar still has a slanted glass cover over a black-painted, water filled basin. As sunlight penetrates the device, solar energy is absorbed by the basin liner and transferred to the water via conduction and convection. Minor heat losses exists from reflection by the glass and water surface, and absorption from the basin liner (energy is transmitted to the ground).

As the water evaporates, water vapor begins collecting on the glass cover. As build-up occurs and condensate beads become larger, gravity overpowers adhesion and the purified water molecules trickle down the slanted glass plate to collect in a gutter designed to capture the pure water and carry it to a storage tank or spigot. Since evaporation is the mechanism of purification, this technology is effective for the complete removal of all chemical, organic, and biological contaminants within the feed water.

However, solar distillation requires higher amounts of solar energy for longer periods of time than does solar pasteurization or even indirect distillation or UV irradiation. Therefore, solar distillation requires the most amount of solar energy compared to the other solar technologies. While the per-volume demand of solar energy may be higher for UV irradiation due to the utilization of photo-voltaic (PV) panels, solar energy captured when the system is not in use can be stored in batteries to supplement the device at a later time, allowing UV irradiation to operate under lower solar flux scenarios than solar distillation.

Additionally, due to the slow rate of evaporation that occurs even on the most ideal day, the production per square meter of the still is low. Because the still is glass covered and tends to be rather large, the capital cost for implementation can be quite high (for manufacturing the glass and delivering it to the site), and the risk of environmental damage is also significant (from animals, weather, and other unforeseen events).

Since there are no moving parts and the only input required is the addition of more water, maintainability of a solar still is extremely simple compared to technologies such as RO, MD, and RBF. Depending on specific construction, slow sand filtration and solar pasteurization may also have similarly low maintenance requirements. In fact, the only maintenance required is to occasionally clean out the basin of contaminants and the removal of algal growth that builds up over time. This is most common when purifying salt water using solar distillation, though cleaning would still be required if contaminated water had only bacteria and protozoa as the dead microbes would eventually form a layer which would begin interfering with the efficiency of the basin.

Solar distillation is a technology that may be readily accepted in rural areas due to its simplicity and smaller scale of operation. Understanding the concept of evaporation and condensation can be easily grasped by anyone, and small, low yield examples can be delivered to villages. As the community witnesses the cupful of pure water produced each day, it will be understandable for there to be a general desire for larger stills to produce more pure water. Perhaps most advantageous in this regard would be to use solar distillation as an entry point to getting wider local acceptance of technologies such as membrane distillation and solar pasteurization which are harder to observe, but produce more water per unit area.

2.7 Solar Pasteurization

Pasteurization is a concept that has been widely accepted for a substantial amount of time for use in purifying milk and other products, but is only recently coming of interest for water treatment. While some health education efforts in the past have encouraged users to boil water to ensure purity, most protozoa and bacteria are inactivated at much lower temperatures. In the table below (also found in Chap. 3) the temperature required to kill contaminants in a given time is shown. The D-value represents the time required to inactivate 90% of the given contaminant, and a value of 5D represents inactivation of 99,999% of microbes or contaminants.

Temperature	Result
55°C (131°F)	Worms, protozoa cysts D value = ~1 min
60°C (140°F)	*E. coli,* rotavirus, *Salmonella typhi, Vibrio cholerae, Shigella* sp. D value = ~1 min
65°C (149°F)	Hepatitis A virus D value = ~1 min

It could be suggested that the rule-of-thumb of boiling water has been given to provide users with a visual metric of ensuring that water has been sufficiently heated, but the additional energy required to bring water to a boil (past the pasteurization temperature) is excessive, particularly for environments where access to fuel is limited. Alternative methods of ensuring that feed water is adequately heated (but not boiled) are mentioned in the chapter on solar pasteurization.

Understanding the temperature range of solar pasteurization technology, it becomes quite simple to explain the purpose of the technology: to heat water to the pasteurization temperature (often taken as 65°C) and no higher to minimize the required energy input. The mechanism used to bring water to this temperature varies and will be briefly mentioned in the following paragraphs.

Regardless of the configuration of solar pasteurization and the metric used to determine when pasteurization temperatures have been reached, this technology uses less energy than any other solar technology mentioned. Other than gravity fed membrane systems and slow sand filtration systems, solar pasteurization systems consumes the least amount of energy per volume output of technologies considered. However, similar to solar distillation and slow sand filtration, the volume purified per unit area is quite low, on the order of magnitude of 0.1 m^3/m^2 as further discussed in the chapter on this topic. However, Duff and Hodgson (1999) found that their systems could produce 100 l/day with a flat panel collector with an area of 0.33 m^2 resulting in a nominal production rate of 0.3 m^3/m^2/day.

2.7.1 Flat Panel Collectors

Practically speaking, flat panel collectors work and look similar to photovoltaic collectors. As with many of the developed sustainable solar technologies, the name declares the function. Water is moved through a flat and often rectangular structure, such that sunlight passes through one transparent side and heats up the water to the pasteurization temperature. The design is such that the ratio between surface area and volume is maximized, therefore, the water in the basin is often shallow.

Surprisingly, these panels are excessively prone to the sun's radiation, and therefore have the capacity to purify water at a higher rate than several other technologies mentioned. The average rate of purification for flat panel collectors is 0.17 m^3/m^2/day, and because of this flat panel collectors are convenient to use in regions with high sunlight.

In fact, by using a heat exchanger built in to the flat panel, Stevens et al. (1998) found an increased flow-rate. In his report, Stevens calculates flow on a per hour basis, referencing 10–55 l/h m^2, which would be 0.24–1.32 m^3/m^2/day, assuming a 24 h solar day. Since pasteurization happens for about 6 h a day, the figure becomes 0.04–0.33 m^3/m^2/solar day which is still potentially higher than most other solar technologies discussed which usually hover around 0.1–0.15 m^3/m^2/day.

One major drawback is that large, flat sheets of material are needed for this system to operate properly. The risk developed by the implementation of large-area, exposed components is significant. Complications from the weather, wildlife, cattle, children, and other variables are significant for this technology. Flat panel collectors have a threat from human damage due to playing or climbing on it or accidentally hitting it with other objects. Particularly if there is an integrated heat exchanger, these non-designed human interactions can have a significant negative impact on the performance of the device. Therefore, the flat panels would require some fencing or security to prevent un-intentional uses.

Fortunately, however, there is a low level of technical knowledge required for this technology, making it easy to operate and easy to train users in operation. Since the primary active components of the system are flat-black absorbent paint and glazing to trap solar radiation, the frame and casing can be constructed of a wide variety of materials locally available to the people. This also helps mitigate the amount of technical knowledge required for construction. Since there are no moving parts, maintenance is quite low, and the only expensive component that may need to be replaced is the glazing over the top of the device. However, this is an expensive component and can be difficult to maneuver to the project site safely.

Unless the device is constructed such that a thermostatic valve is in place or some other mechanical/electrical system exists to ensure water is pasteurized before flowing through the system, the technology would have to be monitored intensively during the hours prior to solar noon when the water is being heated, and after solar noon when the solar flux capacity is dwindling so that water stops flowing when it is no longer hot enough to be pasteurized. This calls for one level of complexity beyond the simple flat panel collector, but is necessary to ensure that water is pasteurized. Since it is a flow-through device, an indicator like the WAPI (SCI 2009) used in solar cookers would be insufficient.

2.7.2 Compound Parabolic Collectors

The technical transition between a flat panel collector and a compound parabolic collector is simple: just imagine holding a piece of paper flat in your hand and folding it into a parabolic shape. Many materials used as reflectors have the capacity to be gently folded without rupturing, and therefore much floor space can be saved by utilizing a parabolic collector. Additionally, higher concentration ratio leads to less square footage and materials required for same production compared to flat panel collectors. Since the concentration ratio is higher, water tends to be pasteurized faster, though in a smaller volume, so that the total production for each solar day remains relatively the same.

Due to the deployable nature of the shape, a wide variety of materials can be used. This is advantageous in seeking locally available resources to use as reflectors. Often cardboard covered in aluminum foil is sufficient in this application, again due to higher concentration ratios. Since there is a small volume of water in the device at any given time, the structure typically weighs less than other technologies discussed, and therefore does not demand such a robust frame, which also lends itself to a wider range of creative design with local materials. Also, as water flows through a tube instead of across a large flat area, the risks imposed by external factors are minimized in comparison to flat panel collectors.

The greatest drawback to this technology is two-fold. First, for high efficiency, parabolic collectors rely on double-walled vacuum tubing, which must be manufactured, and therefore imported. This is a major drawback when considering how locally friendly all other components are for this technology. Second, since it is a flow-through device, it is also difficult to ensure that water was sufficiently heated while passing through the device without advanced temperature monitoring systems. Several solutions have been developed to overcome this obstacle, as outlined in Chap. 3.

One method to determine pasteurization temperatures of effluent water is to use a thermostatic valve from an automobile. This is advantageous for communities that have vehicles and therefore would have access to spare parts. However, as a community becomes more remote, it would become increasingly difficult to find spare parts to repair the valve were it to be damaged. Another method that is being used is to create a disparity in the hydraulic gradient line, such that water of insufficient temperature (and therefore insufficient density) could not overcome the vertical barrier. However, once water was heated to the desired pasteurization temperature, it would expand enough due to a change in density to spill over the barrier and into the storage container. This also presents challenges, as the vertical height of the barrier (most commonly a length of vertical tubing) must be precisely calculated based on the physical properties of water. Therefore, if the tubes were to be damaged it would require a high level of technical competence to reconstruct the design, or it would require a location of safekeeping for the original design plans. Heat exchangers can also be used in parabolic collectors after pasteurized water passes through some temperature check valve.

The most expensive component of this technology is either the temperature valve check or the vacuum tubing, depending on initial design criteria. That said, the cumulative cost of this design is relatively cheaper than all other technologies, except for slow sand filtration. This is particularly true if the design allows for local materials to be used for the reflectors. Parabolic compound collectors are a cheap and efficient technology if solar pasteurization is viable for a given community.

2.7.3 UV Irradiation

Irradiation again takes what has been publicly taught on water treatment and moves a step beyond. In the same way that solar pasteurization doesn't require boiling for water to be purified, UV irradiation does not require heat input for water treatment. Instead this technology relies on applying light at specific wavelengths to contaminated water to deactivate bacteria and protozoa. This process has been found to be highly effective and inexpensive while producing a reasonable volume of water per unit area UV light source. There are two mechanisms used in UV irradiation (or solar disinfection). One is referred to as "direct disinfection" and utilizes the sun's natural wavelengths during sunshine hours to disinfect contaminated water. The other, "indirect disinfection," harnesses the sun's energy via PV panels, and applies very specific wavelengths to contaminated water under UV lamps. Since direct irradiation can be coupled with solar pasteurization as the sun will be heating the water even as it is applying neutralizing wavelengths to the microbes, only indirect disinfection will be discussed here as a unique technology. The most commonly accepted UV wavelength used in disinfection is 254 nm, which comprises only 5% of the total wavelengths emitted by the sun. Therefore, it is much more efficient to develop fabricated UV environments which narrow the UV spectrum to emit only that which is ideal for disinfection (Kim et al. 2008).

Indirect irradiation utilizes various lamps to accomplish its goal of deactivation depending on scope and variety of microbes involved. The common classification of UV lamps is low-pressure monochromatic lamps, medium-pressure polychromatic lamps (with both visible and UV wavelengths) and recently there has been the introduction of pulsed UV lamps (Bohrerova et al. 2008).

First to discuss is some basic knowledge of which wavelengths deactivate microbes and how. In the ultraviolet (UV) range, there exists three sub-groups: UVA,

UVB, and UVC, and are differentiated by wavelength, with UVA being the longest wavelengths (320–400 nm) and UVC being the shortest (100–280 nm). Each bacteria and protozoa is most sensitive to specific wavelengths, therefore it is usually preferable to have some variance in the spectrum of emitted wavelengths, though systems can be tailored easily to common contaminants to save on energy consumption and inventory variance. Low-pressure lamps are optimized at 253.7 nm, as this wavelength has been determined to be optimal for the inactivation of the widest range of microbes (Bohrerova et al. 2008). Medium pressure lamps and pulse-lamps have an acceptable range between 200 and 300 nm, but additionally medium pressure lamps have "long-pass" frequencies, which include the visible spectrum. Therefore, low-pressure lamps are often categorized as using UVB irradiation, medium pressure lamps tend to be classified as UVA, and pulse-lamps are UVC. In the same way that medium pressure lamps have "long-pass" frequencies, pulse-lamps have a shorter wavelength frequency, as low as 100 nm.

The energy required to operate the lamps varies significantly by project, manufacturer, wavelength specificity, turbidity, and purified water production flow rates. In a direct comparison experiment performed by Bohrerova et al. (2008), the energy required to purify water having *E. coli* concentrations of 1×10^8 cells/ml was given. For low pressure lamps and medium pressure lamps, four 15 W lamps and one 1 kW lamp was used, respectively, while pulse-lamp used had an average power output of 2.5 kW. Each pulse lasted only milliseconds, but was triggered every 10 sec. It was found that the continuous wave lamps (low pressure and medium pressure) required 2 sec–2 min to purify water, whereas the pulse lamp required 1–20 pulses to achieve purification (Bohrerova et al. 2008).

It has been found that solar disinfection is an effective technology in the removal of all biological microorganisms and protozoa, with log reductions ranging between 1.05 and 4.26 depending on the average wavelength of light used (Bohrerova 2008). However, it was also found in a different study (Tranvik and Bertilsson 2001) that the effect of irradiation was dependent on the chlorophyll levels in biologic matter in the water samples, with photobleached organic matter having virtually no response to UV irradiation, but organic matter with some coloration responding favorably to UV irradiation.

One of the disadvantages of solar disinfection is the lack of *observability* associated with this technology. Due to the risk of environmental damage to the lamps (weather, livestock, or human interference), often times lamps must be somehow enclosed. Even still, were the process to be observed it is difficult to visually see the treatment process. This phenomenon of wavelength neutralization is much more technically advanced than other theoretical concepts involved in other technologies, and therefore educating users on how and why UV irradiation works will be a more difficult challenge. Because of this technological gap in knowledge, and because it cannot be intuitively deduced that purification happens when non-visible wavelengths are passed through contaminated water, it will be more difficult for UV irradiation to be accepted by communities as a functional technology. More than the other technologies discussed, UV irradiation can come across as a "black box" technology—contaminants go in one side and come out the other side pure, and there is little understood as to why. While this technology has been used in places like Korea and Japan, for many parts of the world such a technology would require significant explanation.

Therefore, while UV irradiation may be on par with the energy per unit output and the area per unit volume requirements of other technologies discussed, it may be a late-adoption technology simply because of the more advanced technical knowledge required to understand the system. Maintenance knowledge is quite low as changing a light bulb is the primary maintenance requirement, but knowing why a specific type (wattage, wavelength, etc.) of light bulb needs to be there requires significant understanding. For those without a true knowledge of the workings of UV irradiation, it may be falsely assumed that a bulb that produces a lot of light purifies water the best. That, and other similarly related misconceptions can easily form around solar disinfection and must be actively educated out of the paradigm of users.

2.8 Technology Development Challenges

Challenges to development of technology could be in terms of added research and development. For water treatment, techniques for effective removal of emerging contaminants, synthetic chemicals, and pesticides, as well as dealing with spills of chemicals in navigable rivers as well as the development of sustainable treatment methods are some of the challenges. It is the issue of sustainable treatment methods that will primarily be discussed here. A great challenge involving technological development is the need to develop technology that is appropriate, relevant, and sustainable. In regards to sustainable water technologies, development and implementation must make economic and social sense to the stakeholders. Technology implementation that provides safe and affordable drinking water—a crucial human need—can markedly improve the human condition for billions across the globe.

Drinking water treatment technologies have been used and continuously developed over the ages. Greek and Sanskrit writings gave suggestions of ways to treat water as long ago as 2000 BC. They thought perhaps heating water would help, and knew that sand and gravel filtration helped to decontaminate water. However, turbidity was the main criterion used in determining purity, as the knowledge of microorganisms was well beyond their time. The earliest known treatment method was used by the Egyptians around 1500 BC, where they applied chemical alum to contaminated water to remove suspended solids. This is now known as the principle of coagulation (Lenntech 2009). Drawings of this system are found in the tombs of the Egyptian Pharaohs Amenophis II and Ramses II.

Sir Francis Bacon, an English philosopher and scientist was the first to attempt salt water desalination in early 1600, and while his attempts did not work, it opened the door for future endeavors to flourish. It was also in this same time period when Dutch scientists first discovered microorganisms, changing the purpose of water treatment drastically. In India, charcoal was first used as a drinking water treatment technology in the seventeenth century as well. Since then, technologies and techniques have been developed as our understanding of science has deepened. Chemical applications of water treatment (like chlorine filtration) weren't discovered until the nineteenth century, and membrane distillation wasn't discovered until the twentieth century.

The need for continued water treatment development was discussed in the introduction, but can also be highlighted here. The average American living in the United States consumes 185 gallons of water per day. If this number is expanded to include all industrial and agricultural demands, then each US resident uses an average of 1,400 gallons/day. This translates into 255 m^3/capita/year for domestic usage and 1,932 m^3/capita/year for gross consumption including all sectors.

The United Arab Emirates (UAE) withdraws and produces the most potable water of the many countries discussed in Chap. 6 on Solar Distallation. However, for comparison, the UAE has access to only 68% of the treated water that the United States uses per year. As such, many countries do not withdraw nearly enough water to support the same level of economic demand or household consumption as in the United States.

2.9 Technological Implementation—Case Studies

In light of the many challenges presented throughout this book, it is important to remember that drinking water treatment technology has been overall successful so far in relieving water stress for millions (in fact, billions) of people. Some selected examples of technology implementation are given below:

- Natural Filtration—Haridwar and Nainital, India
- Membrane Filtration—Singapore
- Solar Distillation—Mexico/United States border
- Solar Pasteurization—Nyanza Province, Kenya

2.9.1 Natural Filtration

Riverbank and a combination or lake/riverbank filtration are presented as examples of natural filtration. Two case studies are presented:

2.9.1.1 Haridwar, India

Due to religious practices, high population, economic considerations, and low enforcement of best management practices for sanitation, the people of India often draw surface water from polluted sources. This is most evident of those in lower economic strata.

The Uttarakhand state of India is located in the northern regions of the country, nestled against China and Nepal. The region of Uttarakhand is the origin of many of the two major rivers of India, the Ganga and the Yamuna from the snow-capped Himalayas. According to Hindu mythology, there are numerous sacred places along these rivers and their tributaries. People bathe at these sacred places. Yet despite the many rivers which originate in the nearby mountains and the region's relatively low population density of 198 people/km^2 (513 people/mi^2), many of the open water sources are unsafe to drink.

The City of Haridwar is located on the River Ganga where it descends from the mountains to the plains. It is one of the most religious pilgrimage places of India with a permanent population of about 200,000. For religious occasions, an additional 330,000 people come to the city (Sandhu et al. 2010). Uttarakhand Jal Sansthan (UJS) is the agency that supplies water to Haridwar. RBF wells located along the River Ganga or Upper Ganga Canal provide 35% of water to the city. The water supply system in the city was developed in 1927 during the British rule of India. At the end of 2009, there were 16 large-diameter bottom entry wells and 31 vertical wells providing about 64 million liters of water per day. Twelve of these 16 large diameter wells are located on a stretch of about 3.3 km along a narrow strip of land between the River Ganga and Upper Ganga Canal and the spacing among them varies between approximately 200 and 300 m. The distance of these wells to the water body (the canal or the river) varies between 3 and 115 m (Sandhu et al. 2010). Four more large diameter wells are located on an island where the Upper Ganga Canal originated near the Bhimgoda Barrage on the River Ganga. The depths of these wells vary from 7 to 10 m below ground surface. The aquifer is relatively shallow with a thickness of 17 m maximum. The production rate from each well varies between 600 and 2,800 l/min.

A number of water quality studies were conducted since 2005 and Sandhu et al. (2010) summarizes them. Sampling between December 2005 and March 2006 revealed that the dissolved organic carbon (DOC) of the bank filtrate was less than 1 mg/l and the arsenic content was less than 0.01 mg/l. The systems operated under aerobic conditions. All trace metals were below the Indian drinking water standards. The source water of the River Ganga was also low in DOC (0.6–1.2 mg/l) at Haridwar. Dash et al. (2010) showed 2.5 log removal of total coliforms, 3.5 log removal of fecal coliforms, and 0.7 log removal of turbidity during non-monsoon periods (November to June) with a total travel time varying between 84 and 126 days. A log reduction of 1 represents 90% removal of matter tested. During monsoon periods, the river has high turbidity, flow, and microbial contaminants. The log removals for total and fecal coliforms as well as for turbidity were 4.7, 4.4, and 2.5, respectively. During the monsoon period, the travel time of water varied between 77 and 126 days. The abstracted water from the RBF wells at Haridwar is only disinfected (primarily by chlorine) before supply.

2.9.1.2 Nainital, India

While rivers are numerous, the topographic challenges do not always permit easy access to these water sources, and therefore many people rely on the various lakes in the Uttarakhand region. The town of Nainital used to rely on Nainital Lake for its water supply. Recently, bank filtration has been used as the primary source of drinking water supply.

In a case study done in Nainital by the Indian Institute of Technology (Dash et al. 2008) seven pumping wells were developed less than 100 m from the edge of Nainital Lake to test the effectiveness of bank filtration between 1997 and 2006. It was determined that the wells were impacted by groundwater infiltration to the lake, lake seepage, and RBF from one of the perennial inlet drains. This multiple effect scenario was a product of the monsoon season, and therefore the study analyzed water quality in both the monsoon and non-monsoon seasons and compared the two. Lake contamination is caused primarily by the city's antiquated leaky sewage system coupled with direct disposal of sewage into the lake. In an attempt to mitigate these sanitation issues, quick sand filters were used as far back as 1955. However, these filters have since been proven inadequate to match the city's demands or meet the national water quality standards. In order to produce potable water from the polluted lake, advanced treatment methods will be needed. However, the resources to build, operate, and maintain an advanced water treatment plant are not available for many residents in the area. Therefore, natural purification using bank filtration was considered as the best solution to the problem.

Due to the stratification of the unconsolidated detritus material the wells were bored into, it was determined that the water-bearing strata reached a depth of 36 m. Therefore, all seven wells were between 22 and 36 m in depth. Using Darcy's law, the scientists calculated the travel time between the lake and the wells. The monsoon season would yield the shortest travel time as it reflects the highest volume of water. For the wells that were close to the lake shore, it was determined that the water only remained in the ground for 1–2 days (the further wells—84 m from the shore—were calculated to take 11–19 days). This is significantly lower than the assumed value of time required for proper coliform removal during natural filtration, which has been reported to be 10–20 days (Medema et al. 2000).

Despite the short filtration time, and despite the fluctuations in water content during various seasons, it was determined that water arriving at the tube wells had achieved a 4–5 log reduction in total coliform, and a 1.6 log reduction in turbidity. Additionally, suspended solids, bacteria, chemical oxygen demand (COD), and chlorophyll-a were all reduced below detectable limits. The removal of contaminant was so efficient that the water was good for drinking without any further treatment. Conversely, the quick sand filtration used prior to the natural filtration resulted in a coliform count of 2,300 MPN/100 ml, well above the quality standard of a maximum 50 MPN/100 ml coliform concentration. Therefore, the sand filtration has been abandoned in the Nainital region and they have constructed two additional pumping wells to provide clean water to the city of over 50,000 people through the use of bank filtration. The availability of clean water has increased significantly in the region, as the new wells produce 24 Ml/day of pure water where the old sand filters produced only 1 Ml/day of inadequate water.

Therefore, it can be seen that the use of natural filtration is viable and effective for regions with access to alluvial deposits hydraulically connected to a surface water source. It is additionally seen that the use of natural filtration provides consistently high quality water in larger quantities than mechanical filtration despite a significantly longer travel time through the filtering medium.

2.9.2 *Membrane Filtration in Singapore*

Due to the worldwide shortages of water and growing demand for fresh water supplies, the process of desalinating water has gained more attention. As of now, "Worldwide membrane and thermal desalination capacity is over 11 billion gallons per day from over 12 thousand plants, worth $ 9.2 billion per year, growing at a rate of 12% per year" (AMTA 1–2). Particularly for Singapore, an island nation with limited fresh water and abundant seawater, desalination is an attractive option. Desalinated water has recently found many uses throughout the world including industrial, power plant, military, touristic, and most notable municipal. The Singapore government recently showed its confidence in this technology and that of solar energy by allocating $ 170 million towards its research and development. Scientists are currently working on integrating the membrane distillation processes along with solar, geothermal energy, and heat waste to develop cost efficient energy saving processes for desalinating water.

Major disadvantages of membrane distillation are low productivity and high costs. Researchers in Singapore are working to improve the flux in the process and modify the current techniques. They are using a series of systematic module configurations in an attempt to enhance the total flux. These configurations are made of designs including the baffle, external/inner helix, and can be sieved during module fabrication. They have also brought in two very unique configurations to the module: spacer and twisted modules.

> By implementing these different module designs, the investigators observed an 11–49% increase in flux performance at 75°C with respect to the original, unaltered module. It is interesting that the highest flux attained (49% increase) combined two plastic sieves and the inner helix configuration at 75°C. The generation of turbulence, the increase in effective membrane-surface contact, and the effects of cross-flow possibly account for the improvement in MD performance. (Teoh et al. 2009)

These new module configurations are not only proving to be beneficial, but are sparking interest in both the Middle East and United States for further development.

In reference to the cost issues regarding seawater distillation, the past few decades have shown a significant decrease in the pricing of desalting elements. Due to technological advances, competition, and automation of suppliers worldwide, seawater membrane costs have visibly reduced. "In the last decade, desalting technology has improved significantly and costs have decreased by over 50%" (AMTA 2007). The validity of membrane distillation with the development of new technologies in Singapore and supplier's cost reduction has dramatically improved the technology's feasibility. Membrane distillation's ultimate capabilities may prove to be the wave of the future and if we look to Singapore as an example, we see that the membrane distillation theories discussed in this book are not only relevant, but extremely necessary. We need to diminish our water shortages and fight the increasing fresh water demands while efficaciously utilizing the Earth's resources.

2.9.3 Solar Distillation—Mexico/United States Border

In many dry regions around the world, the lack and growing need for drinking water directly correlates to high solar insulation. In particular, the arid communities along the US-Mexico border face water supply issues that are comparable to those faced by the developing world. As a result, a need for a low cost effective solution to provide safe drinking water is a necessity. The adoption of solar distillation technology could provide clean water for some developing countries worldwide. As an example, in many border cities including Chihuahua (Ciudad Juárez County), New Mexico (Doña Ana County), and Texas (El Paso County) this technology has been used. In these three counties alone, there are over half a million people with very limited water supply and minimal infrastructure development. Municipal water supplies are not up to par with national drinking standards and contain contaminants and high levels of arsenic. "Distillation is the only stand alone point-of-use (POU) technology with NSF (National Sanitation Foundation) certification for arsenic removal, under Standard 62" (Foster et al. 2005).

To acquire safe drinking water, these residents purchased potable water from elsewhere at premium prices and hauled it back home. Solar distillation offers an attainable, on the spot solution to give clean water access to these border families. Over the last decade, solar distillation technologies have been developed in these cities to demonstrate the practicality of this technique along the border and ultimately throughout the developing world. "EPSEA worked closely with NMSU during this initial pilot demonstration, where 40 pilot $3' \times 8'$ solar distillers were built by EPSEA and distributed to colonia families and health clinics in West Texas." Studies showed that these solar stills efficiently removed all salts, heavy metals, biological contaminants (*E. coli*, *Cryptosporidium*) and water borne pathogens from contaminated water sources in addition to some pesticides due to the UV rays, high temperatures, etc. "Average water production is about 0.8 l per square meter per sun hour." Daily solar still production for a square meter for two days can be seen in Fig. 2.2.

Progress has continued with this technology and the effectiveness of these solar stills has granted them much attention and financial support.

> Two grants were awarded by the Border Partners in Action (BorderPACT) with the Consortium for North American Higher Education Collaboration to disseminate solar stills distributed to 27 Mexican families in Chihuahua. EPSEA also won a community challenge grant from EPA to distribute stills to 80 families in Texas and New Mexico from 2000 to 2002. (Foster et al. 2005)

These solar stills work by the simple concept of evaporation and condensation, which were described in more detail earlier. After the water has evaporated, all

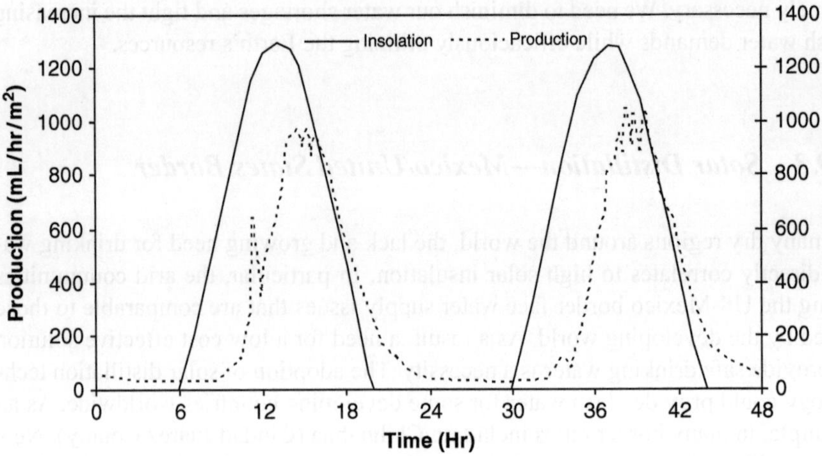

Fig. 2.2 Solar insolation and measured solar distillate production over 48 h in southern New Mexico; notice how distillate production lags insolation and continues even after sunset. (After Foster et al. 2005)

contaminates are left behind and the evaporated water is not only purified, but desalinated as well.

> Typical production efficiencies for single basin solar stills on the Border are about 60% in the summer and 50% during the colder winter. Single basin stills generally produce about 0.8 l per sun hour per square meter. (Foster et al. 2005)

Still design has significantly improved over the last few years. What began as a silicone membrane lined still in the mid 1990s, converted to aluminum to avoid beading of the water droplets, and is now comprised of ABS Plastic by a company coined SolAqua. Outgassing issues were addressed with new proprietary inner membrane materials and the distillation technology has shown significant progress. "The technology has now evolved to the point where with large manufacturing volumes unit costs could be greatly reduced by a factor of three or more in the future" (Foster et al. 2005).

Additionally user response has been quite positive. Most found it cheaper to purchase a low priced solar still rather than traveling to purchase bottled water at high prices. "Solar still savings were approximately $ 150–200 a year per household instead of purchasing bottled water" (Foster et al. 2005).

> Many families in the U.S. colonias (border communities) often spend from $ 8 to 12 per week on bottled water. Likewise, in northern Mexico families often spend $ 3–5 per week on purified water. This represents an investment of anywhere from $ 150 to 600 per year for bottled water. (Foster et al. 2005)

Thus the payback period of a still versus bottled water is only 2–3 years, with savings amounting to thousands of dollars over a decade. "The levelized energy cost of solar distilled water is about US$ 0.03 per liter, assuming a ten year still lifetime" (Foster et al. 2005).

In some cases, owners highly valued the idea of a clean, affordable technology that left a minimal carbon footprint on the Earth. A few drawbacks on the technology however, are that the stills are not producing optimal amounts to meet production needs in the winter.

> Generally, it appears that for most Border households about 0.5 m^2 of solar still is needed per person to meet potable water needs consistently throughout the year. Those households with insufficient wintertime still water production typically had 0.35 m^2 or less of still area per person. (Foster et al. 2005)

Only about 40% of users are receiving sufficient water production all year round. (Foster et al. 2005). However, supplemental water cost is still far less than residents having to purchase bottled water throughout the summer when premiums are at their highest. As of now, EPSEA has expanded its distribution to and began solar stills implementation in Australia, the South Pacific, Mexico, and Guatemala. It is evident that this technology and these solar stills allow a practical, relatively inexpensive way for residents to obtain drinking water. As seen in these overall successful borderland city cases, solar stills have astounding worldwide potential to address potable water needs and ultimately, saving lives.

2.9.4 Solar Pasteurization—Nyanza Province, Kenya

As was mentioned earlier, 1.1 billion people do not have access to safe drinking water and much of Africa's population contributes to this astounding statistic. About half of the population of sub-Saharan Africa does not have access to clean water. Although there are insufficient water sources in most regions of Africa, there is an abundance of sunlight. This excess sunlight can be transformed into the energy required for solar water pasteurization through the use of solar cookers. Ultimately, this solar energy can heat water to temperatures that kill harmful microbes and provide safe, clean drinking water for the people.

> What is not well known is that contaminated water can be pasteurized at temperatures well below boiling…Used alone, boiling and solar [pasteurization] were about twice as effective as chlorine [disinfection], and when used together they were four times as effective. (SCI 2009).

Currently, there is a Sunny Solutions program which began in Africa in 2003, being implemented in the Nyakach region, Nyanza Province, in western Kenya. As a part of this program, women can choose to use the CooKit Solar Cooker to prepare food and decontaminate their own water. In this particular area of Kenya, there is a very high occurrence of typhoid fever as well as bacterial and amoebic dysentery. Additionally, their wells and streams are highly contaminated with *E. coli*. However, with the use of the CooKit solar cookers, there has been a substantial decrease in diarrheal diseases and many other water borne diseases primarily caused by *E. coli*.

> The Nyakach solar cooks and village leaders have been taught how to use innovative water testing methods, Colilert tubes and Petrifilms to test their water before, and after solar pasteurization. The package of water testing materials, CooKit, and WAPI, combine to address two main problems in developing countries: lack of wood for cooking and unsafe water for drinking. (Metcalf 2009)

Both the Colilert (www.palintest.com) and Petrifilm (http://solutions.3m.com/wps/portal/3M/en_US/Microbiology/) water sampling methods enable people to accurately and cheaply test water. These methods provide quality microbiology and give water testing abilities to the 72 districts of Kenya.

Results of these women and leaders properly utilizing and benefiting from the Sunny Solutions proposition seems to show these solar cookers being the solution to making clean water and safe food available to the poorest of the poor. "When… I visited the homes of 16 [CooKit users] in July 2004, we found that each woman was heating water in a CooKit when she was not cooking, and was pasteurizing 5–10 l/day." A survey taken in mid-July of 2005 indicated that amongst the 47 Kenyan families chosen, solar pasteurization was quickly adopted and the use of the CooKits provided a visible reduction in diarrheal contamination among small children (Metcalf 2009). While pasteurization has been accepted in the food industries for decades, it has been seemingly left out of discussions regarding unsafe water.

> From published data and our own experiments, we established that heating contaminated water to 65°C will pasteurize the water and make it safe to drink. (Metcalf 2009).

The success of the solar pasteurization projects in Kenyan validates the potential for this type of technology and reflects highly upon its need in developing countries. Solar energy is often the only energy source for people living in remote areas in developing countries. Solar pasteurization can thus provide safe drinking water in most cases where water does not contain contaminants such as inorganic chemical pollutants or arsenic.

Chapter 3
Solar Pasteurization

Ed Pejack

Abstract Solar pasteurization is one of the easiest and cheapest methods to produce potable water in remote areas. It relies on the energy of the sun to heat the water to pasteurization temperature. Most people believe that water needs to be boiled to kill pathogens present in water. However, pasteurization temperature is sufficient to kill these microorganisms. The process is similar to conventional milk pasteurization process with the exceptions that the source of energy is the sun and the use of low cost materials. Besides providing potable water, solar pasteurization can be used for cooking. Solar pasteurization finds application in remote, high elevation sunny areas where electricity is not available. In these areas, firewood is not easily available. This chapter provides some of the basics of solar pasteurization, presents easy to find materials that can be used to fabricate the pasteurization systems, and compares the performance of commercial and easy to make units. Its use for potable water production is discussed.

Keywords Solar pasteurization • Potable water • Pathogens

3.1 Microbiology of Water Pasteurization

Diseases caused by waterborne pathogens affect a large number of people, principally in the developing world. Several billion people use water that is likely contaminated, resulting in approximately 2.5 billion cases of illness per year, and about 5 million deaths per year (Burch and Thomas 1998).

Heating of water to a sufficiently high temperature for a certain time period destroys harmful microorganisms, which is often termed, 'pasteurization' after the scientist Louis Pasteur (1822–1895). The required temperature and corresponding time period for destruction of microbes are shown in Table 3.1 (Ciochetti and Met-

E. Pejack (✉)
School of Engineering and Computer Science, University of the Pacific,
3601 Pacific Avenue, Stockton, CA 95211, USA
e-mail: epejack@pacific.edu

C. Ray, R. Jain (eds.), *Drinking Water Treatment,* Strategies for Sustainability,
DOI 10.1007/978-94-007-1104-4_3,

Table 3.1 Effect of temperature on microbes

Temperature	Microbes destroyed (D~1 min)
55°C (131°F)	Worms, protozoa cysts
60°C (140°F)	*E. coli*, rotavirus, *Salmonella typhi, Vibrio cholerae, Shigella* sp.
65°C (149°F)	Hepatitis A virus

calf 1984). The D-value is the time to kill 90% of the organisms. A time of 5D would kill 99.999% of the microbes.

At temperatures higher than those listed in the table above, the D value decreases significantly; e.g., for the microbes in Table 3.1 with D value of ~1 min at 60°C, the D value decreases to ~12 s at 65°C. Also, considering that water as it heats toward 65°C also spends time at 64°C, 63°C, etc. (at which temperatures addition fractions of microbes are killed), it is widely taken that solar pasteurizing to 65°C will make it safe to drink. Many pasteurization devices use a safety factor of several degrees above 65°C to allow for variations in measurement and equipment.

The time required to pasteurize decreases exponentially with temperature (Feachem et al. 1983), as illustrated in the semi-log plot of time vs. temperature shown in Fig. 3.1.

It is a common misconception that water pasteurization requires boiling of water (sometimes even stated as for 20 min!). Unnecessary boiling would waste significant fuel when it is already an expensive or scarce item in much of the world. Boiling requires about twice the energy as heating to 65°C, plus the extra time of monitoring, and the time and expense of obtaining fuel. Conceivably the boiling

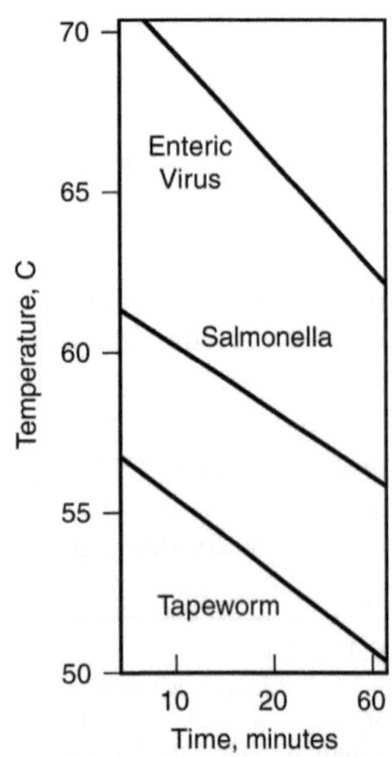

Fig. 3.1 The semi-log plot of time vs. temperature required to pasteurize. (From Feachem et al. 1983)

criterion may have arisen because it is often visibly evident when water boils, and then one is certain that 65°C has been exceeded.

Note that the pasteurization discussed in this chapter is concerned with the destruction of microbes only. Toxic chemicals, metal salts, and other contaminants are not removed. Insoluble material such as silt can be decreased by pre-filtering or settling.

Pasteurization as discussed in this chapter is a thermal process, relying on the destruction of microbes by temperature (achieved with solar energy). A related but different water treatment concept is solar disinfection (acronym SODIS), where microbes are destroyed by direct action of certain wavelengths of the solar spectrum, independent of temperature. It is the ultraviolet range of the spectrum (200–400 nm) that is more effective in destroying microbes; consequently it is desired to employ transparent containers that have good transmittance in this wavelength band. The interested reader can find many reports of studies and trials of SODIS, for example, Caslake et al. (2004) and Meierhoffer and Wegelin (2002).

3.2 Use of Solar Cookers for Drinking Water Production

There is a widely disseminated body of knowledge and literature on the subject of solar cooking and solar cookers, and several international conferences have been held on this subject in the past 20 years and archived by Solar Cookers, International (Sacramento, California). Since many foods are cooked in the range of 80–100°C, it is not surprising that solar cookers, containing only water, are also used to pasteurize at 65°C.

Perhaps the first type of practical solar cooker, the "box cooker" is commonly used around the globe. In its simplest form it is a box insulated on sides and bottom with a transparent top cover and frequently a "booster reflector" at the top to augment the solar power entering the top cover (Fig. 3.2a).

There have been numerous analyses and clever design attributes of box cookers, developed in recent decades, which will not be delved into here. Some of the relevant design and operating parameters are:

- Thermal insulation of box
- Optical properties of top cover
- Surface characteristic of box interior
- Booster reflectors
- Solar power, cloudiness
- Time (hour, day, month)
- Latitude
- Device orientation, tracking/non-tracking
- Mass and type of food
- Pot size, surface, lid
- Wind, ambient temperature

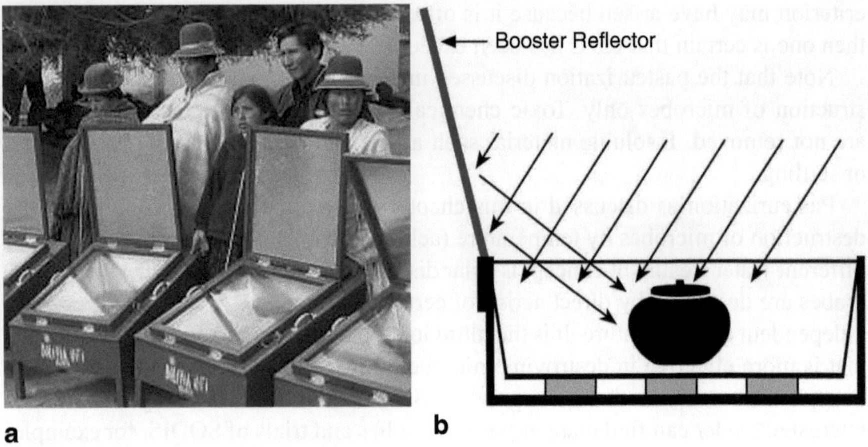

Fig. 3.2 The box cooker: **a** wooden box cookers with slanted cover, made in Bolivia; and **b** cross-sectional schematic drawing of a box cooker

A great variety of box cookers are in use around the world; some are small and hold one pot, others range in sizes holding over ten pots. Box cookers (Fig. 3.2) have been made from cardboard (one of the first being the famous Kerr-Cole cooker), wood, metal, plastic or mud. Thermal insulation for the box is necessary, and again very many materials have been used, including crumpled paper, air space, paper baffles, feathers, rice hulls, and foam.

The simplest application of a box cooker to pasteurizing water is to place a pot of water in the cooker as in cooking food. It often occurs that a reasonably designed box (able to cook a variety of foods in 2 h) with a solar radiation of 700 W/m^2 or so can pasteurize a liter of water in about an hour.

Another popular cooker design, termed a "panel cooker," consists of several flat reflective surfaces (panels) adjacent to a pot, which is enclosed in a transparent bag or enclosure. There is no insulated box; rather the heat loss from the pot is minimized by the layer of hot air between the pot and the bag. Commonly used bags have been made of polypropylene film, but other transparent films have been used (such as polyethylene, nylon). The bag should not fit tightly on the pot, so as to avoid melting from the hot pot and to avoid too much heat loss. If the bag has too large an air space, the insulating ability of the air space is decreased. An air gap of around 1–2 cm is recommended. The ideal material for the bag would have a high transmittance for solar radiation wavelengths (0.3–1 nm) and low transmittance for thermal radiation wavelengths corresponding to radiation at around 100°C (5–14 nm). Two types of panel cookers are shown in Fig. 3.3.

The "panels" are reflecting surfaces that vary in number and orientation. The panel's function is to reflect solar rays onto the bag, increasing the concentration and solar power to the bag (and the water pot). Numbers of panels range from one to a dozen or so, and are usually flat planes to take advantage of the ease of construction and foldability; however curved, hemispherical, and other modified panels have been employed.

Fig. 3.3 **a** The CooKit (SCI 2009a) with pot inside plastic bag, and **b** the "Hot Pot" has a glass bowl instead of a bag. There is an air space between the inner black pot and the bowl. (Photo credit: Patricia McArdle)

A particular version of panel cooker, termed "the CooKit" (Fig. 3.3), developed by Solar Cookers International (SCI 2009a) has the desirable attributes of low cost and weight, ease of construction, and foldability. It has seven flat panels when unfolded. The one front flap is adjusted upward to reflect on the pot or bag. One useful characteristic is that it does not need to be re-aimed at the sun for several hours, as the panels compensate for the relative solar motion. As with the box, the pot can be filled with polluted water and, with sufficient solar radiation and time, pasteurizes it with approximately the same time as the box cooker.

A large number of panel cooker variations have been developed and produced in recent years. Variations include number and shape of panels, overall size, type of container, material of construction, and method of folding/unfolding.

A third class of solar cooker where the design emphasis is to concentrate the solar radiation much more than a box or panel cooker, is termed "concentrators." This type employs parabolic surfaces (simple or compound), Fresnel lenses, or multiple aimed reflectors. Heat rates are much higher than for box or panel cookers. Users must take care not to get their arms or face into the concentrated solar rays. Since the radiation to the pot is so high, a bag or cover around the pot is usually unnecessary. To utilize the reflective character of parabolic surfaces, concentrators should be tracked to the sun more often than box or panel cookers. One example of a concentrator is shown in Fig. 3.4.

3.3 Devices Designed Specifically for Water

A simple model of the heat balance of a pasteurizing device maybe considered as follows (Fig. 3.5).

Assume a constant solar power Q_s (watts) entering a pasteurizing device where the heat loss to ambient at temperature T_0 (Celsius) is characterized by a constant heat loss coefficient U (watts/degree Celsius), where the heat loss power is

Fig. 3.4 Example of a concentrator. (SCI 2009b)

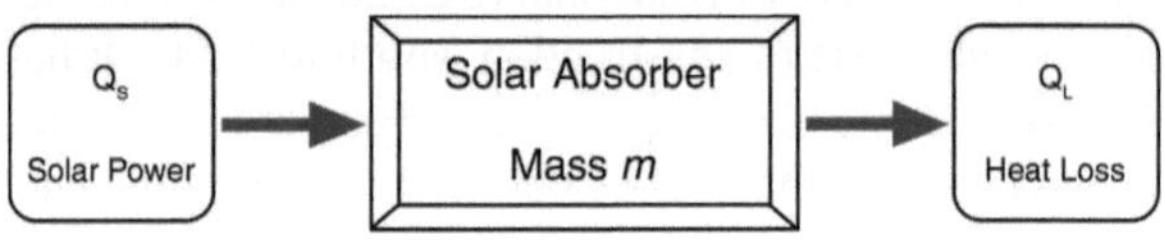

Fig. 3.5 Energy model for a simple solar pasteurizer

$U(T-T_0)$. Assume that the initial temperature of the polluted water is the same as ambient, T_0. The differential equation for heating, after integrating from initial time to time t_p (the time to reach pasteurizing temperature, T_p), becomes

$$t_p = \frac{-mC_p(T_p - T_0)\ln(1-\lambda)}{Q_s\lambda} \tag{3.1}$$

where C_p is the specific heat of water, and a new dimensionless parameter λ, the heat loss ratio, is defined as the ratio of the heat loss power at temperature T_p divided by the solar power Q_s.

$$\lambda = U\frac{T_p - T_0}{Q_s} \tag{3.2}$$

For the limiting case of zero heat loss ($U=0$), the terms involving λ in Eq. (3.1) have the value -1, and the time to heat to T_p equals a more familiar expression

$$t_p = \frac{mC_p(T_p - T_0)}{Q_s}. \tag{3.3}$$

From Eq. (3.3), some generalizations for effective pasteurizers (minimum time to pasteurize) can be observed. The time to pasteurize is decreased by:

- Increasing solar input power Q_s, by operating when the solar power is highest (e.g., around solar noon, or a sunny location), or using reflectors to achieve concentration.
- Decreasing the mass of water in the device. If the initial mass is too great, pasteurization temperature may not be achieved before the solar flux becomes too low, later in the day.
- Decreasing the initial temperature difference between pasteurization temperature (e.g., 65°C) and initial temperature of the polluted water. Starting with warmer water decreases the time to pasteurize.
- Decreasing the heat loss coefficient, U, or the heat loss ratio, λ. Minimize heat loss from the water as it heats by using thermal insulation, paying attention to thermal radiation from the device, and avoiding a windy location.

This simple model of a pasteurizing device neglects the effect of water evaporation. If water vapor is allowed to escape the device, this can represent a huge energy loss, increasing the time to pasteurize. As noted above, minimizing the heat loss factor decreases the time to pasteurize. A very low loss factor can be achieved by vacuum insulation, which practically eliminates heat loss by convection. One of several such systems (Fig. 3.6) is the Solar Kettle (Kee 2006), where the water is contained in a glass vacuum insulated tube 0.75 m in length. The outer surface of the inner tube is coated with a selective surface (such as aluminum nitrate as demonstrated in Fig. 3.6), enhancing absorbance of solar rays and minimizing re-radiation of thermal radiation.

Fig. 3.6 An array of Solar Kettles articulated to pour out the pasteurized water. (Photo credit: Alex Kee)

3.4 Simple Devices from Common Materials

Surprisingly simple devices made from common materials can pasteurize small quantities of water in 30 min to 1 h. A discussion here of simple, common, inexpensive pasteurization devices is made because it is often the case that poor, distressed peoples do not have safe water, neither do they have the resources to employ sophisticated, complex water treatment or devices. Perhaps the simplest pasteurization device is a simple flat puddle (Pejack et al. 1996; Andreatta 2009) of water with a transparent cover and exposed to sunlight, with or without adjacent planar reflectors.

Experimental results (see Table 3.2) with this simple flat geometry were obtained when using a transparent polypropylene bag to contain the water and zero, one or two adjacent vertical reflectors each 0.30 m×0.30 m (1 ft×1 ft) of aluminum foil. When two reflectors were used, they were oriented 90° to each other with the water between the reflectors. The size of the flat water puddle was 0.25 m×0.41 m (10 in.×16 in.), and the puddle rested on a foam surface to minimize heat loss from the bottom. The experiments were performed in June in Stockton, California on a sunny day when the noon solar flux was 980 W/m^2.

Andreatta (2009) describes an amazingly simple method of implementing the flat puddle pasteurizing method, called the "sack pasteurizer." Starting with a circular sheet of transparent plastic about 1 m diameter, and the circumference is bunched up to form a sack to hold 3 l of water and tied shut. The sack was placed on various surfaces (black plastic on grass, foam or bubble wrap; grass alone, black foam), which puts the water into a somewhat flat puddle, and then covered with a sheet of transparent plastic similar to the bag. An air gap is made between the bag and the cover. Experiments at 40° N latitude (Ohio, USA) in late August on a strong sunny day achieved pasteurization temperatures.

Another simple water pasteurizer can be made with commonly obtained cylindrical bottles. The configuration consists of two vertical reflective surfaces of aluminum foil oriented 90° to each other, plus a third reflector on the ground (Figs. 3.7 and 3.8). The total reflector area was 0.36 m^2 (3.9 ft^2). The water containers used were:

Bottle 1	0.36 l glass, spray painted black
Bottle 2	1.0 l soda (polyethylene terephthalate [PET]), painted black
Bottle 3	0.5 l brown glass beer bottle

Table 3.2 Test results with a simple solar puddle

Time to pasteurize (min)	Water depth (cm)	Water volume (l)	Reflectors
25	0.40	0.41	Two
79	0.98	1.00	Two
54	0.98	1.00	One
109	0.98	1.00	Zero

Fig. 3.7 A black 1 l PET bottle inside a cover made from two 2 l PET bottles (Bottle 2, Cover 2); the reflector in front is tilted up. (Photo credit: Ed Pejack)

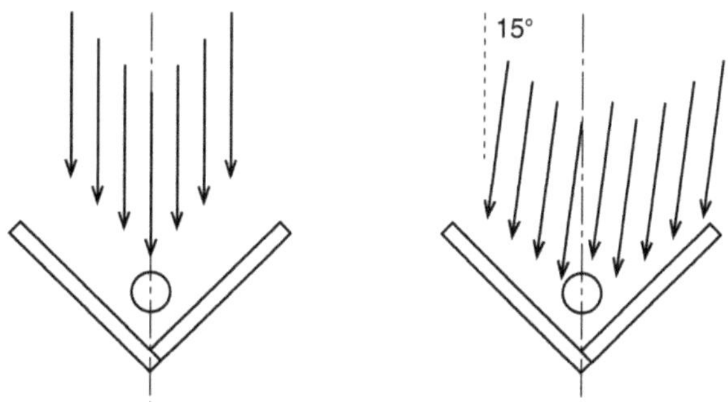

Fig. 3.8 Bottle pasteurizer with included angle of 90° vertical reflector planes. At *left* solar rays are on centerline, at *right* with 15° azimuth

Transparent covers for the bottles were made in these ways:

Cover 1	polypropylene bag
Cover 2	2 l PET bottles slit around the middle, then one bottom half slit six times lengthwise and the other bottom half slid over the other
Cover 3	no cover

Results for experiments (on the same day as the flat geometry discussed above) are:

Time to pasteurize (min)	Container
116	Bottle 1, Cover 3
82	Bottle 2, Cover 2
72	Bottle 2, Cover 1
51	Bottle 3, Cover 1

Sometimes the question is raised concerning whether exposure of PET plastic bottles to sunlight could cause plasticizers to enter the water. Studies have indicated that solar disinfection devices using PET are safe with respect to human exposure to plasticizers di(2-ethylhexyl)adipate (DEHA) and di(2-ethylhexyl)phthalate (DEHP) (Yegian and Andreatta 1996). It is suggested that the safe use of PET or other plastic containers be investigated with respect to potential health effects.

The simplicity of the notion of using a cylindrical water container between two vertical rectangular reflectors stimulated a ray-tracing study (by the author) of the effects of included angle between reflectors and azimuth angle of the entering solar rays. When the solar rays bisect the angle between the reflector planes it is referred to as zero azimuth. The included angle of 90° was found to give higher concentration than 0, 45, 60, or 180° between the panels. Somewhat counter intuitive, for 90° included angle, the concentration for an azimuth of 15° was higher (3.6) than the concentration for zero azimuth (3.0). These two cases are shown in Fig. 3.8. Consequently, the two 90° planes could be oriented with azimuth of 15°; then for 2 h the relative solar movement would result in concentration decreasing from 3.6 to 3.0 for the first hour, then increasing from 3.0 back to 3.6 for the second hour, without any adjustment of the reflectors. In the case shown, the water cylinder was placed two diameters from the intersection of the two reflectors.

These studies suggest that 0.5–1 l of water may be pasteurized in 1–1.5 h using simple material and simple geometry. It occurs too often that large numbers of people such as refugees, victims of catastrophes, etc. are congested in an area with contaminated drinking water. In those situations such methods described above may be very advantageous, as well as for travelers in remote regions where safe water is unobtainable.

Many variations of simple pasteurization devices have been developed using various configurations of water containers, covers, and reflectors (Yegian and Andreatta 1996; Cengel 1998; Duff and Hodgson 1999).

3.5 Commercial Devices in Production

A form of the puddle type pasteurizer described above is commercially produced as AquaPak® (Fig. 3.9), which is a small (0.35 m×0.35 m) flat polyethylene plastic bag with one side transparent bubble pack insulation, and the other side black. The pasteurizing ability depends on ambient temperature, solar power, and the quantity of water (2–5 l). It is reported to have been used in 30 countries. When the solar radiation was 800 W/m^2 and initial water temperature was 25°C, it was reported that 2 l of water were pasteurized in 2.2 h. The lid has a small wax filled capsule that indicates (by displacement of the melted wax) whether or not the water has been pasteurized. Water pasteurization devices are discussed later. A larger capacity system, the SunRay 30 (Fig. 3.9) is a flat collector (79 cm×61 cm×9 cm) containing ten black bottles. It was reported to pasteurize 7.5 l in 1.5 h.

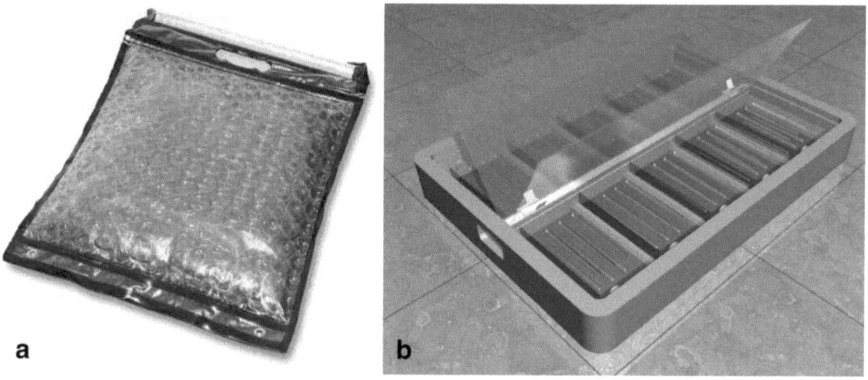

Fig. 3.9 Commercialized pasteurizers **a** AquaPak (Photo credit: Solar Solutions), and **b** SunRay 30. (Photo credit: Safe Water Systems)

3.6 Devices with Recovery Heat Exchange

In the devices discussed above, pasteurized water is either removed at pasteurization temperature, or allowed to cool in the device for use later. The energy in the pasteurized water, once used to heat to pasteurization temperature, can be transferred to water not yet pasteurized, thereby requiring less solar energy. Manifestation of this notion employs a heat exchanger and makes the device a flow-through device, as opposed to a batch process device. In a flow-through device, water continuously flows into and out of the device as pasteurization proceeds. The savings in energy or the increase in quantity of pasteurized water can be quite significant as will be shown in the following analysis. A model of the system is shown in Fig. 3.10.

The flow-through pasteurizer works as follows: as shown in Fig. 3.10, polluted water at ambient temperature T_0 enters one side of the heat exchanger at mass flow rate $\dot{m}$, and exits the heat exchanger at higher temperature T_1 where it enters the solar device. In the solar device, solar power Q_s enters and some loss power Q_L is lost to ambient. The flowing water achieves pasteurization temperature T_p in the device, exits the device and enters the other leg of the heat exchanger. The water flows counter-currently, transferring heat to the incoming stream, finally exiting the heat exchanger at temperature T_2, which is lower than T_p but slightly higher than T_0.

It is assumed for this analysis that the solar power input Q_s, $\dot{m}$ and T_0 are constant. The heat loss power Q_L is here characterized by a loss coefficient U defined such that U times the temperature difference $T_p - T_0$ gives the heat loss Q_s (watts).

We see that the device has only to heat water from T_1 to T_p, requiring less solar power than if heating from T_0 to T_p. The heat balance on the device gives

$$Q_s - U(T_p - T_0) = \dot{m}C_p(T_p - T_1) \tag{3.4}$$

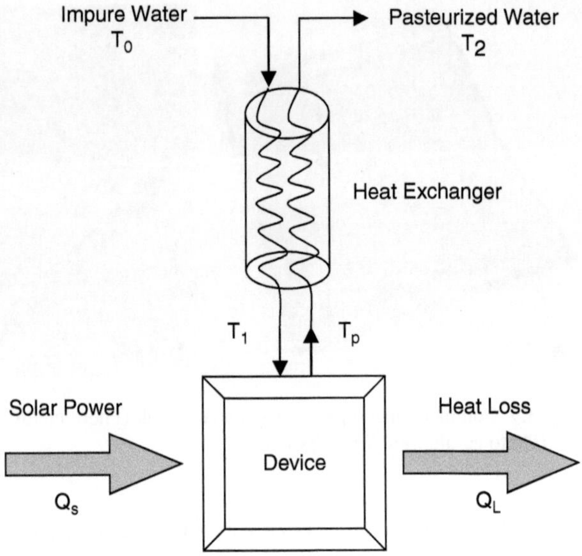

Fig. 3.10 Essential features of a flow-through pasteurizer

where C_p is the specific heat of water. The heat exchanger is characterized by an effectiveness, ε, defined as

$$\varepsilon = \frac{T_1 - T_0}{T_p - T_0} \tag{3.5}$$

for the special case here when the same mass flow rate and specific heat exist on both sides of the heat exchanger. The effectiveness of any heat exchanger depends on mass flow rate, fluid properties and the fluid flow characteristics within the flow passages.

Using the heat loss ratio as defined earlier in Eq. (3.2), combining Eqs. (3.4) and (3.5), and solving for $\dot{m}$, resulted in

$$\dot{m} = \frac{(1-\lambda)}{(1-\varepsilon)}\frac{Q_s}{C_p}(T_p - T_0). \tag{3.6}$$

The reader may observe and be surprised that for heat exchanger effectiveness approaching unity, the mass flow rate of pasteurized water gets very large. This impossible limiting case of $\varepsilon = 1$ implies perfect heat exchange, meaning all of the heat in the pasteurized water is transferred to the polluted water, requiring no solar input once the device is operating at steady state. It is noted from Eq. (3.6), however, that higher heat exchanger effectiveness (ε) and lower energy loss from the solar absorbing device (λ) increases the possible flow rate of pasteurized water. A reader interested in heat exchanger theory, design and analysis may consult any of the many references on the subject (e.g., Cengel 1998). Other analyses of pasteurizers with heat exchangers can be found in Stevens (1998), and Duff and Hodgson (2005).

Comparing flow-through pasteurizers with the batch process type pasteurizers, it is noted that whereas the batch process type functions with a fixed mass of water, the mass flow rate in the flow process can be adjusted in operation to account for the variable solar power. The batch process in most cases uses a fixed mass of water, which if started with too much mass of water, or if the solar power decreases during heating, the entire batch may not become pasteurized. Likewise, if the solar power is more than expected, the mass of water pasteurized would be less than was possible with a larger starting mass.

The flow-through pasteurizer often uses a thermostatic valve to control the water flow rate. A thermostatic valve opens and closes in response to temperature. Some applications have adapted the thermostat valve from the cooling system of automobiles, with the notion that the valve would allow flow only when the temperature is above pasteurization temperature.

The valves operate from closed to open over a temperature range, making the mass flow rate vary with temperature, opening more at the higher temperatures and closing more at the lower end of the range, sometimes with cyclic operation.

Duff and Hodgson (1999, 2005) have designed and operated several versions of a heat exchanger pasteurization system in which the flow is controlled by the density difference between hot (pasteurized) water and colder (unpasteurized) water. Heated water is directed to a vertical riser tube, and when the water there is hot enough, the lower density (and higher water column) allows the pasteurized water to "spill over" and flow through the heat exchanger. The system eliminates the operation problems with a thermostatic valve. Some of the performance data reported includes an average hourly production rate of pasteurized water of 16.4 kg/h (1 kg water = 1 l water) for a 2 h period when the normal radiation averaged 913 W/m^2 on a mostly sunny day. The production of pasteurized water for the entire same day was 86 l, producing pasteurized water from 8:00 AM to 3:00 PM. Their collector was a set of five evacuated heat pipes with a total area of 0.45 m^2. The heat pipes have a relatively low heat loss coefficient.

Many heat exchangers are of the concentric tube or flat geometry type. Yegian and Andreatta (1996) discuss experiences with a concentric tube heat exchanger consisting of an inner and outer annulus separated by a copper tube. The outer annulus contains wrapped wire to enhance the heat transfer. Also discussed is a flat geometry heat exchanger made with sandwich-like layers of copper sheet, and wood. Rubber strips force the water to take a serpentine flow path to increase heat transfer. Heat transfer effectiveness values for the two designs were found to be 60–80%, depending on water mass flow rate. Mass of water pasteurized compared to the batch process was reported to be an improvement by a factor of eight.

It is interesting to compare the mass of water pasteurized with a flow-through device to the mass pasteurized with a batch process, for the same amount of time and the same solar and loss power.

Fig. 3.11 A commercial pasteurizer, the SunRay 1,000. (Photo credit: Safe Water Systems)

For the batch process, a mass m of water is pasteurized in time t_p. Assuming for simplicity that the loss power is again based on temperature difference $T_p - T_0$, then mass m, and t_p are related as

$$m = \frac{(1-\lambda)Q_s}{T_p - T_0} t_p. \tag{3.7}$$

We assign $\dot{m}\ t_p$ as the mass pasteurized in the flow-through device in time t_p, then the ratio μ defined as the mass pasteurized in the flow-through device to the mass pasteurized in batch process, is

$$\mu = \frac{1}{1-\varepsilon}. \tag{3.8}$$

Equation (3.8) indicates that the mass of water pasteurized can be greatly increased, compared to the batch process, with very efficient heat exchangers. A system designed for flow-through operation is shown in Fig. 3.11, and has an exposed area of 3.2 m×1.8 m. It was reported to pasteurize 1,000 l/day (1 m^3/day), which results in the capacity to pasteurize 0.17 m^3/m^2/day.

A system could be assembled using a solar collector originally designed for service water (e.g., for heating buildings, etc.) and adapting a counterflow heat exchanger.

3.7 Water Pasteurization Indicators

A useful attribute of many batch process solar water pasteurization devices is that once they are positioned in the sun, they may be left unattended. However, suppose that a device is left alone in strong sun, and the user returns late in the day. Had the water, now cool, reached pasteurization temperature during the day? Unless a recording thermometer had been installed, the temperature time history of the water would be unknown. It is only necessary to know if the water had achieved the critical pasteurization temperature during the day. What is needed is a sensor that can indicate, and preserve that indication, whether or not pasteurization temperature had been achieved during the day. Such an indicator could be based on a number of physical phenomena, such as

1. Melting of a material from solid to liquid
   A material that melts at the pasteurization temperature could be arranged so that once melted, it changes shape or location and maintains that change when the temperature later decreases.
2. Differential thermal expansion
   Two materials with different coefficients of thermal expansion could be made to interact so that a detectable change in geometry is produced at pasteurization temperature, and that change is able to be maintained when temperature is later decreased.
3. Color change
   Certain materials experience color change at specific temperatures. If the material changes color at certain temperatures, that could indicate pasteurization.

One must be aware that in a volume of water, there will likely be temperature variations.

Temperatures at the bottom of a container may be several degrees Celsius colder than at the top. Therefore the temperature indicator should be placed at lower levels of the water. In unusual cases where the heat input to the container is from the bottom (as for example in a concentrator), the spatial temperature variation may be quite different. In any particular situation, spatial temperature variations should be measured and the indicators placed at the colder regions.

A device based on phenomenon 1 above has been widely used in a variety of solar pasteurization devices, in several forms. The WAPI (Water Pasteurization Indicator) (SCI 2009) is comprised of a small quantity of wax (soybean wax, for example, melting at ~70°C) inside a plastic (or glass) tube sealed at both ends as shown in Fig. 3.12. Initially the wax is all at one end of the tube. If the WAPI is suspended vertically with the wax end on top, then as the wax melting temperature is achieved, the wax melts and moves by gravity to the lower end of the tube. Thus, finding the wax at the bottom indicates that the water has been pasteurized, even after the water has cooled. The WAPI can easily be reset for further use by inverting and using again. It is used with a flexible string and a weight so that it may be suspended in water in vertical orientation. The WAPI is used in many locations worldwide.

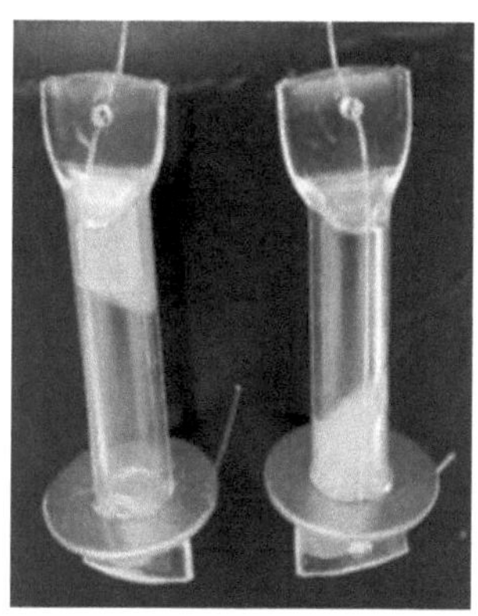

Fig. 3.12 Two views of a WAPI (Water Pasteurizer Indicator). The view on *left* has the wax at *top*. After reaching melting temperature, the wax migrates to the *bottom*, as shown on the *right* view. The metal washer slides on the tube so as to keep the tube vertically suspended by the string. (Photo credit: SCI 2009)

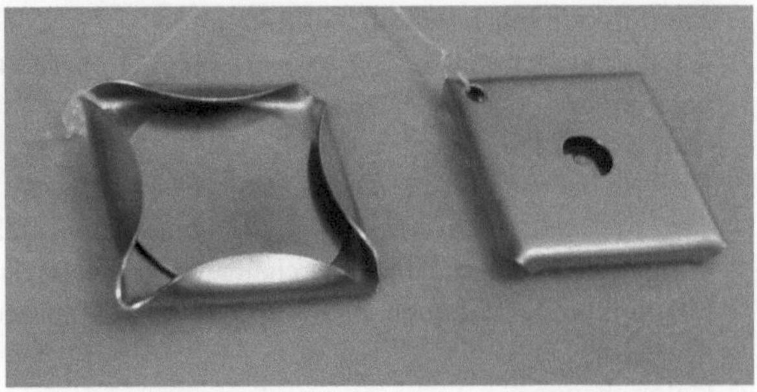

Fig. 3.13 Two Snap Disks with folded housing (~2.5 cm^2). (Photo credit: Ed Pejack)

A water temperature indicator based on differential thermal expansion of dissimilar metals shown in Fig. 3.13 has been termed the Snap Disk (Saye and Pejack 1994). It is constructed of two dissimilar stainless steel disks pressed together and bonded in the form of a 2.54 cm disk. At any temperature the disk has a concavity on one side and is convex on the other. As the temperature is raised and reaches the snap temperature, the disk snaps into reverse concavity (an example of what is called "snap buckling" in mechanics of materials). The new snapped position remains so, even when the temperature decreases again below the snap temperature. The snap temperature is quite precise and repeatable over millions of cycles. The manufacture of snap disks is done in many countries and is a relatively mature technology used in temperature control devices. Disks can be made with any snap temperature and tolerance.

To use the snap disk, one snaps the disk to the starting position, and it is placed in the water container. Once the water is solar heated to the snap temperature, it automatically snaps to the opposite side. Then at any time later, when the water has been cooled, the user can observe if the disk had been snapped and the water pasteurized, because the disk does not snap back when cooled. To reuse the disk, the user physically pushes the disk with the fingers to snap it back to the starting condition. The disk is fast acting, and if lowered into water hotter than the snap temperature, it immediately snaps with a distinct audible "pop."

The user of course must be able to distinguish the two sides of the disk. In the Snap Disk shown, the disk was placed in a folded stainless steel housing open on one side and having a hole on the other side, for manually snapping the disk, making the two sides of the disk readily discernable. The housing also prevents the disk from being overstressed when being manually reset. The disk had a snap temperature of 68°C with a manufacturer's tolerance of 2.8°C, but tests on samples showed the tolerance to be much less. In actual use in water devices, after a time the Snap Disk shown would experience some corrosion from being exposed to various water chemistry, and because of the intimate contact of dissimilar metals. Therefore the disk should be dried when not in use. The manufacturer reported that other metal combinations could be used in the disk to provide better corrosion resistance.

3.8 Multi-use Systems

In addition to the solar water pasteurization applications (~65°C) described above, other uses for solar thermal devices include:

- Autoclaving for medical sterilization (>122°C)
- Producing distilled water for medical use
- Producing service hot water (50°C)
- High temperature steaming and cooking (>150°C)

A multi-use system designed for several of the uses above can be envisaged. Such a multi-use system utilizing solar collectors, heat exchangers, and associated valves and piping conceivably could be feasible or practical compared to design and construction of individual systems for the uses given above. The higher temperatures would likely require low loss collectors (e.g., vacuum tube collectors) and solar concentrators, and the design and operation would be considerably more complex and expensive than simple pasteurization devices.

3.9 Summary

The nature of a solar pasteurization system can take on many forms, as seen from the above discussion, and whatever is appropriate, practical, and feasible for any particular application would require some amount of thought, analysis and planning.

When contemplating and initiating a solar water pasteurization system, the choice arises as to the trade-off between using commercially available devices or adapting raw materials to the construction of custom devices using the basic concepts outlined in this chapter. Commercial solar pasteurization devices have desirable attributes in that they are more likely to perform as intended, have a history of operation, have a robust design, and have clear operating procedures. However, because of variable local conditions of the end users, even well developed commercial devices need to have some support from the manufacturers. But commercial devices may not be feasible in some cases because of purchase cost, transportation, and availability of spare parts. In those cases a custom design and unique construction may serve the users better as well as being more environmentally appropriate.

An example of implementation of solar pasteurization for sustainable drinking water is the 2008 Safe Water Project in Kenya initiated by Solar Cookers, International (Sacramento, California, USA), directed by Dr. R. Metcalf (SCI 2009a). Among the important issues to consider are the end users, which can be:

- One or two persons (at home, hiking or trekking)
- One family
- An extended family
- A village
- A commercial enterprise

Furthermore, the water system design or concept must be chosen with consideration of:

- Initial cost
- Operating cost
- Installation
- Maintenance
- Complexity of operation/repair
- Availability of materials/parts
- Safety and reliability
- Modularity (increasing capacity)
- Portability (size, weight)

Additionally, implementation of a solar water pasteurization system, taking note of the above challenges, needs to include considerations relating to continued, sustainable operation. Finally, the end users need to have some training to ensure successful operation, repair, and evaluation of the system.

Chapter 4
Membrane Desalination

Kishore Rajagopalan

Abstract Increasing population, industrialization, and the desire to improve the quality of life among the world's poorest are all contributing to the increasing demand for water of adequate quality. The competing demands for water range from ecological services, food and feed production, power generation, and shipping, to domestic and industrial needs. While the earth is indeed a "blue planet," fresh water constitutes less than 2.5% of the available water (Postel et al., Science 271:785–788, 1996). Two-thirds of this water is tied up in glaciers and permanent snow cover leaving barely 1% available to supply the growing demand to sustain and enrich human life. Therefore, measures such as sustainable watershed development and protection, rainwater harvesting, and responsible use of groundwater sources are needed to make access to clean, affordable water a reality. While these measures will help, they will not suffice to prevent water from becoming a critical bottleneck in the development of large parts of the world that are facing population growth, dwindling fresh water supplies, and increasing pollution of existing water supplies. They must be supplemented by actions to identify, develop, and upgrade alternative sources of water to meet the anticipated gaps between the long-term supply and demand. Seawater is the most abundant water resource on the planet covering 71% of the Earth's surface, mostly in oceans and other large water bodies. Water extraction from this salt solution (desalination) would provide a reliable source of fresh water for the foreseeable future. This chapter examines membrane-based desalination technologies with an emphasis on seawater reverse osmosis (RO). Environmental impacts associated with membrane-based desalination and efforts to mitigate them are discussed. Furthermore, the challenges of harnessing this technology to serve the needs of the poorest sections of society and the potential pathways to overcome them are explored.

Keywords Desalination • Reverse osmosis • Sustainability • Appropriate technology development

---

K. Rajagopalan (✉)
Illinois Sustainable Technology Center, 1 Hazelwood Drive, Champaign, IL 61820, USA
e-mail: kishore@istc.illinois.edu

C. Ray, R. Jain (eds.), *Drinking Water Treatment,* Strategies for Sustainability,
DOI 10.1007/978-94-007-1104-4_4, © Springer Science+Business Media B.V. 2011

4.1 Desalination Technologies

Commercially significant desalination technologies can be broadly classified as thermal or membrane-based. Thermal desalination technologies add or remove energy in the form of heat to accomplish the separation of water from saline water. Distillation and freezing are two examples. In membrane-based processes such as reverse osmosis (RO), mechanical energy accomplishes the same objective. In other membrane processes such as electrodialysis, the energy is electrical. A list of desalination technologies is provided in Table 4.1. In addition to these technologies, other types of membrane-based processes have been proposed or are under active research. Two of the most promising, membrane distillation and forward osmosis, are discussed later in this section.

4.2 Thermal Desalination Technologies

4.2.1 Multistage Flash Distillation

Multistage flash distillation (MSF) is by far the most common distillation method used worldwide. A MSF system consists of three main sections: a heat input section; several intermediate stages where heat recovery takes place; and a heat rejection section where excess heat is rejected to the environment (Fig. 4.1). Seawater, preheated through passage in the intermediate stages, is brought to operating tempera-

Table 4.1 List of desalination technologies

Technology	Energy source	Principle	Commercialization stage
Multistage flash distillation	Mostly thermal some electrical	Evaporation/condensation	Currently available
Multiple-effect distillation	Mostly thermal some electrical	Evaporation/condensation	Currently available
Reverse osmosis	Electrical	Separation of water from salt through filtration	Currently available
Electrodialysis	Electrical	Separation of salt from water through ion transfer	Currently available
Membrane distillation	Mostly thermal some electrical	Evaporation/condensation	Pilot stage
Forward osmosis (NH_3-CO_2 based)	Mostly thermal some electrical	Induces movement of water into an osmotic agent (mixture of NH_3-CO_2). Subsequent decomposition of osmotic agent leaves water behind	Laboratory studies/small pilot plant

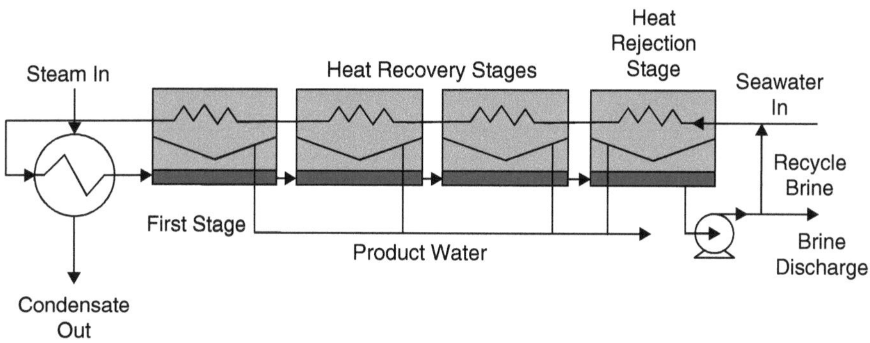

Fig. 4.1 Schematic of a multistage flash distillation process

ture (generally 90–110°C) in the heat input section using steam. It is then routed to the first stage, which is maintained at a pressure lower than the equilibrium pressure of the heated seawater. This results in a portion of the seawater flashing into vapor. The flashed vapor extracts energy from the bulk of the seawater cooling it slightly. Additionally, the removal of flashed vapor increases the salinity and boiling point of the residual seawater. The flashed vapor is passed through demisters that remove entrained droplets and flows into a heat exchanger at the top of the stage. Here, the vapor cools and condenses after exchanging its heat with incoming seawater. This exchange facilitates efficient heat recovery. The condensed water vapor is collected as product water. The residual seawater from the first stage flows to the next stage. To promote additional flashing, the pressure in the second stage must be lower than the first stage to compensate for the drop in temperature and increase in salinity. This process continues until the final stage. In the heat rejection section, the residual brine is cooled to permit discharge to the ocean. Incoming seawater is used for cooling in the heat rejection section. A portion of this cooling water is used as feed to the process.

While MSF is considered a mature technology, advances and improvements are being made (Wade 2001). The long tube distiller which uses an axially placed brine flow system is expected to allow even greater capacities (currently 75,000 m^3/day), lower pressure loss, and higher performance than the current cross flow designs (Al-Sahali and Etouney 2007).

4.2.2 *Multiple-Effect Distillation*

Multiple-effect distillation (MED) is similar to MSF with regard to the input, recovery, and rejection of heat (Fig. 4.2). Key differences lie in the mode of evaporation and the regimes of heat transfer. In MED, the salt water is sprayed on top of horizontal tubes to form a thin film. Water vapor flows internally within the tubes. The transfer of heat from the condensing vapor causes the thin film of water external to

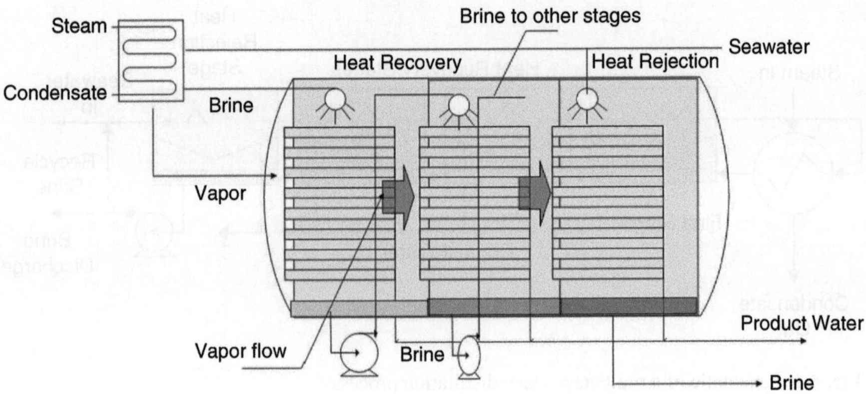

Fig. 4.2 Schematic of a MED plant

the tube to boil. The resulting vapor, after disengagement from entrained droplets, is used as a heat source in a subsequent effect. The heat transfer coefficient achievable in this mode is significantly higher than in MSF enabling reductions in heat transfer area.

Current MED plants typically operate at lower temperatures than MSF. Consequently, chemical scaling, i.e., precipitation of sparingly soluble salts, is less of an issue. In the Low Temperature MED (LT-MED) the operating temperature can be as low as 40–65°C. The benefits of operating at a lower operating temperature include reduced level of pretreatment, potential for use of low grade energy, and use of less expensive materials such as aluminum for heat transfer areas. Recirculation flows are also lower in MED relative to MSF systems. As in MSF, the primary energy source in MED is steam and the cooling medium in the heat rejection stage is seawater. Coupling of MED with thermal or mechanical vapor compression allows for more compact systems.

4.3 Membrane Processes

4.3.1 Reverse Osmosis

In RO, water is extracted by pressurizing saline water through semi-permeable membranes that allow passage of water but not salt (Fig. 4.3). This ability to inhibit the flow of salt is termed salt rejection, defined as the ratio of salt concentration in the filtrate (permeate) to the salt concentration in the feed. A minimum pressure—the osmotic pressure—has to be exceeded to operate the process. The osmotic pressure increases with water salinity. Pressures appreciably exceeding the osmotic pressure are used to ensure an acceptable rate of water production. Production rates are reported in gallons of filtrate per square foot of membrane surface area per day

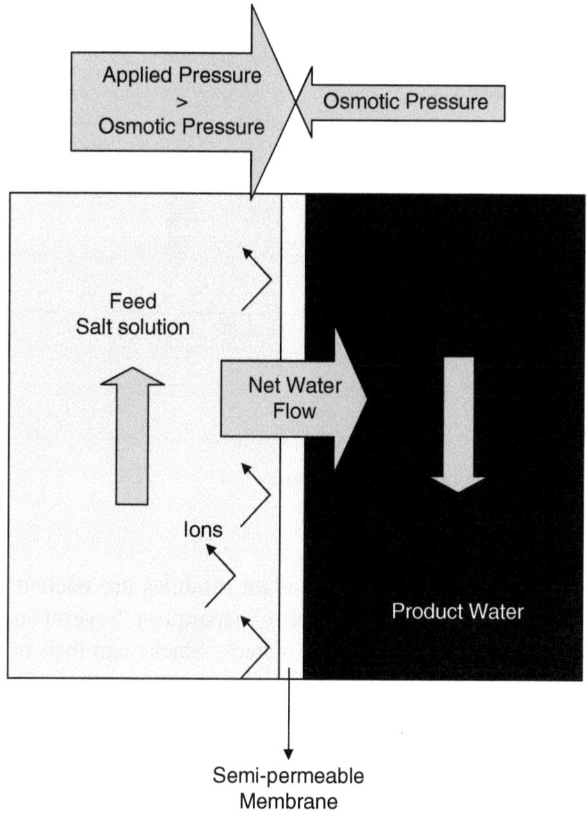

Fig. 4.3 Schematic of a reverse osmosis process

(gfd) or in other units such as liters of filtrate per square meter of membrane surface area per hour (LMH).

In the early years of RO, asymmetric cellulose membranes were used as the semi-permeable barriers (Saltonstall 1977). In these membranes, the separation layer and the underlying support were made of the same material. The development of thin film composite (TFC) membranes in the 1980s was a major development in improving the performance characteristics of RO membranes (Cadotte et al. 1980). Composite membranes consist of a semi-permeable barrier laminated to an underlying support. This allowed the optimization of the membrane layer and the support structure individually resulting in major improvements in membrane permeability and salt rejection characteristics. The TFC's used are generally comprised of aromatic polyamide or polyetherurea films deposited onto a porous support by chemical reaction.

The membranes are manufactured as flat sheets and subsequently rolled into spirals (Fig. 4.4). This configuration, the spiral wound membrane, provides a high ratio of membrane surface area to packing volume. The diameters of spiral wound membranes have steadily increased over the years and today, module diameters as large as 16 in. can be manufactured, each with a membrane area of 160 m^2

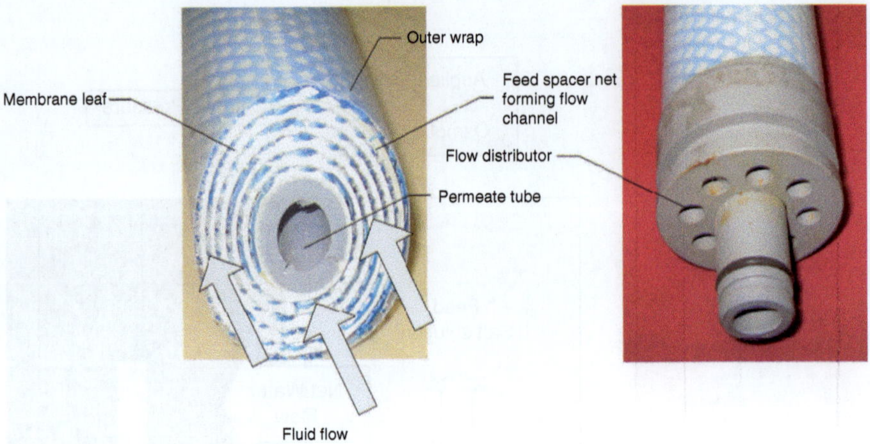

Fig. 4.4 A spiral wound membrane

(Anonymous 2008). Up to eight modules are packed into a fiberglass-reinforced housing and constitute a unit of separation. Several such units are typically linked together in parallel to form a stack. Stacks can then be linked together in series to form a complete system.

Depending on the salinity of the feed, water recovery—the quantity of filtrate produced per unit of feed—can range from 40 to 85%. The lower rate is common in seawater applications while the higher rate is more representative of brackish water (typically water with ionic content between 2,000 and 5,000 mg/l) applications. In many instances, brackish water recoveries are limited more by the likelihood of precipitation of sparingly soluble salts than the salinity of the feed water.

RO process efficiency is highly sensitive to feed water quality. Feed water of poor or marginal quality can lead to chemical scaling, biofouling, and particulate blocking of membrane channels necessitating thorough pretreatment. Feed water treatment for RO may include sand or cartridge filtration, coagulation, flocculation, softening, silica removal, chlorination, dechlorination and the like, depending on the composition of the feed water. The higher tolerance of distillation processes, MSF and MED, to feed water quality variations allows them to compete effectively where energy costs are lower.

4.3.2 Electrodialysis

Though electrodialysis was the first commercialized membrane-based desalination technology, it currently serves a relatively small market. The technology removes cations and anions from the salt water through application of an electric field. A

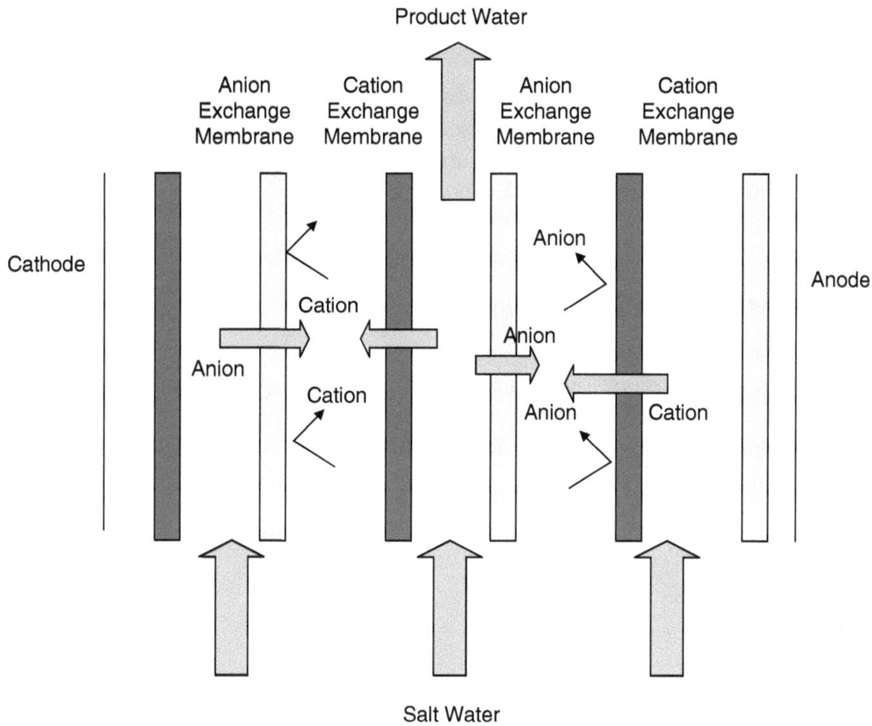

Fig. 4.5 Schematic of an electrodialysis system

stack (Fig. 4.5) is comprised of alternating cation and anion selective membranes placed under an electric field gradient. The end compartments contain the anode and cathode along with their respective electrolyte solutions. Feed water flows in the space between the membranes and is subjected to a potential gradient. Under the influence of the electric field, the cations permeate through the cation selective membrane towards the cathode while the anions migrate towards the anode. As the ion selective membranes are in an alternating arrangement, cations migrating into an adjacent compartment are restricted from further movement towards the cathode by the intervening anion selective membrane and vice versa. The migrating cations and anions are trapped in the intervening channels forming the concentrate compartment. The feed solution in the meantime is depleted of both cations and anions. The energy for separation is proportional to the current flow and increases with salinity. Highly saline waters such as seawater contain many more ions than brackish water and therefore require more energy for separation. To desalinate seawater from 36,000 to 800 mg/l salt may require 26 kWh/m^3 (93.6 MJ/m^3) per one estimate (Korngold 1982). This energy vastly exceeds that of processes such as RO. Hence, electrodialysis is normally used for treating brackish waters of low to medium salinity.

4.4 Emerging Membrane Technologies

Two emerging processes—membrane distillation and forward osmosis—show particular promise for small and medium scale applications. They also have the ability to handle highly saline waters. These are best classified as hybrid technologies as they combine both membrane and evaporative processes.

4.4.1 Membrane Distillation

In membrane distillation (Fig. 4.6), the feed water is brought into contact with a microporous hydrophobic membrane. The hydrophobic nature of the membrane pores prevent the passage of liquid water below a threshold pressure but allow permeation of water vapor.

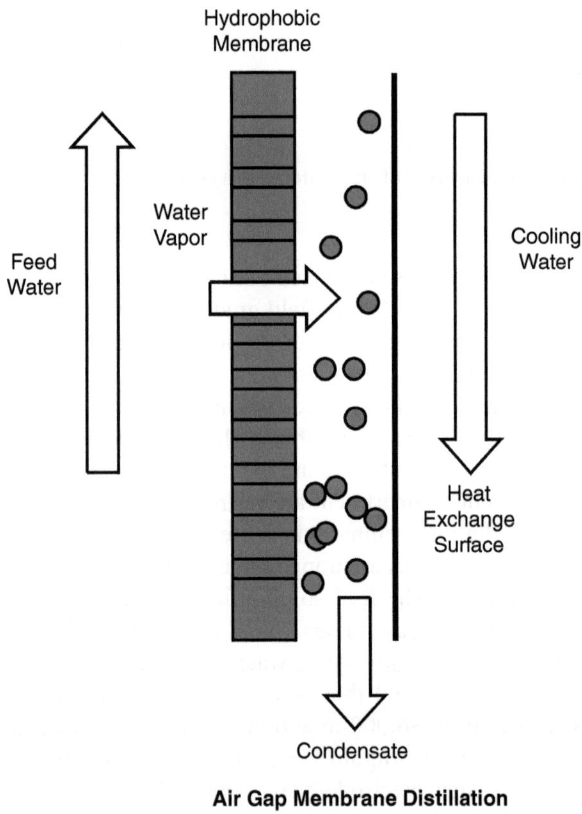

Fig. 4.6 Illustration of an air gap membrane distillation process

A water vapor pressure gradient is generated across the membrane by maintaining a temperature differential: a hot feed and a cold condensate. This promotes the diffusion of water vapor from the hot side to the cold. Salt, being a non-volatile solute, is prevented from passing through resulting in separation. The diffusing vapor condenses on the cold side of the membrane. As the process involves a phase change, latent heat has to be supplied.

Unlike MSF or MED, the extraction of latent heat in membrane distillation is more complicated. One commercial system, the Memstill®, is currently available with integrated heat recovery (Hanemaaijer et al. 2006). The reported specific energy—energy to produce a unit amount of water—consumption is 100–200 MJ/m^3 of distilled water (Dotremont et al. 2008).

4.4.2 *Forward Osmosis*

Forward osmosis refers to the natural movement of water across a semi-permeable barrier from a solution of low osmotic pressure to one of high osmotic pressure (Fig. 4.7). Typically, the saline water is the solution of low osmotic pressure from which water is extracted. The higher osmotic pressure solution—the draw solution—is a solution of organic or inorganic substance.

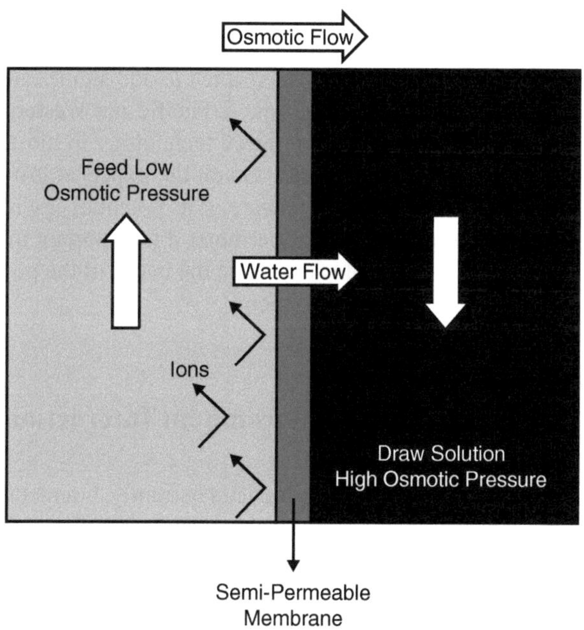

Fig. 4.7 Illustration of a forward osmosis process

In some respects, forward osmosis is equivalent to the proverbial jump from the frying pan into the fire. The theoretical energy required for separation of water increases with an increase in osmotic pressure of the solution. In forward osmosis, as the water is being transferred to a solution of higher osmotic pressure, it implies that a higher minimum energy of separation would be required for subsequent extraction of water from the draw solution. In spite of this argument, there is interest in developing this technology due to the potential use of low-grade energy, low-pressure operation, and the possibility of increased water recovery.

A process using NH_3-CO_2 has been developed that utilizes a solution of ammonium carbonate to extract water from saltwater across semi-permeable membranes (McCutcheon et al. 2005). The diluted ammonium carbonate solution is then heated to release ammonia and carbon dioxide leaving water behind as a product. While the energy required for such decomposition is high, it can be accomplished with low-grade energy. Secondly, the process is also capable of extracting water from solutions with as high as 20% salt content. Electricity requirements are claimed to be low at 0.25 kWh/m^3 (0.9 MJ/m^3) (McGinnis and Elimelech 2007).

4.5 Global Growth of Membrane Desalination

Global desalinated water production capacity is estimated to grow from 39.9 to 97.5 million m^3/day over the period 2006–2015 (Alcolea et al. 2009). The North American market over the same period is expected to grow from 7 to 11 million m^3/day. Other major regions expected to add significant new capacity are the Middle East, North Africa, East Asia & Pacific and Western Europe.

RO is likely to be the preferred technology in most parts of the world with the exception of the Middle East. Given the expected growth in membrane based desalination, it is critical to be aware of its potential environmental impacts and ongoing efforts at mitigation. Furthermore, it is important to recognize the challenges of harnessing this technology to serve the needs of the poorest sections of society and seek pathways to overcome them.

4.6 Desalination Environment Interactions

Desalination processes, as practiced currently, interact with and impact the environment at many levels. The impacts are site specific and dependent on water source and water quality. Impacts result from energy consumption, construction and operation of seawater intake structures, brine discharge, use of water treatment chemicals, discharge of corrosion products, equipment noise, aesthetic and visual impairment of coastal areas and restrictions on fishing.

4.6.1 *Intake Structures*

Seawater intake structures are designed to obtain consistent feed water quality. Therefore, intake structures are usually located away from ports and outfalls to avoid pollution and in areas where seawater characteristics are not unduly influenced by natural events.

Intake structures can, if improperly sited or designed, draw in large quantities of aquatic organisms by impingement or by entrainment. Impingement refers to the trapping of marine organisms against the intake screens due to high velocity water flows. Once pinned, the fish have a low probability of breaking loose and returning to the sea. Many species of fish and juveniles have a 24 h mortality rate upwards of 85%. On the other hand, smaller non-mobile organisms, larvae, and eggs can be swept in, a process termed entrainment. Mortality rates in these cases are usually 100%. Therefore, a careful assessment of the potential impact of impingement/entrainment on reductions in biodiversity or fishing stocks should be carried out before undertaking desalination projects.

4.6.2 *Brine Discharge*

A natural consequence of desalination is the production of an effluent with higher salt concentration—the brine. Table 4.2 lists selected characteristics of desalination brine from the predominantly used technologies of RO and MSF and their potential impacts (Lattemann and Hopner 2008).

Some recent insights on impacts of brine discharge have been obtained based on operation of one of the largest seawater RO desalination plants located at Ashkelon, Israel (Safrai and Zask 2008).

Salinity increases of 3% above ambient were measured at distances of 500 m along the shoreline while 1% increased salinity has been observed for a few kilometers from shore. Monitoring at this test site has also detected a "red brine" phenomenon due to discharge of ferric ions (as high as 42 mg/l) used in the pretreatment process. The discolored plume has been observed at distances as far as 3 km from the outfall and near designated marine reserve areas under certain weather conditions. Because the load of these ferric ions is around 500 t/year, it is speculated that cumulative effects could include reduced primary production rates due to high turbidity and reduced light penetration, creation of an anoxic zone due to sedimentation, accumulation of iron in filter feeding species, algal blooms, changes in diversity of species and the like. Interestingly, these effects were not anticipated during the initial permitting process.

It is clear that as the density of desalination plants increases around coastal areas, the likelihood for interaction between discharges and their cumulative effect will constitute an increasing burden on the receiving waters. The knowledge base for predicting their effect, both in the short and long term, is still sparse and evolving.

Table 4.2 Characteristics and effects of brine discharge from RO and MSF Plants. (Adapted from Lattemann and Hopner 2008)

Brine characteristics	RO	MSF	Effect
Salt concentration	6–7%	5%	Increased salt concentration increases density. This is counteracted by increased temperature. Therefore, RO discharges tend to sink but MSF discharges will either be neutrally or positively buoyant. The effect is a change in local salinity due to brine discharge
Temperature	Ambient	5–15°C above ambient	Increased temperature decreases oxygen content and reduces brine density
Oxygen	Decreased due to use of sodium bisulfite	Low due to deaeration and use of oxygen scavengers	Increased temperature and salt content lowers oxygen solubility in water
Chlorine	Not a concern—chlorine removed prior to contact with membrane	0.2–0.5 mg/l; shock dosing in the range of 6–8 mg/l	Adverse impact of chlorine and chlorinated by-products on organisms
Halogenated organics	Can be present in RO concentrate	Varying composition and concentration	Recommended limit—long term 7.5 μg/l, short term 13 μg/l; Halogenated organics have been detected in proximity of distillation plants; no hard evidence of adverse effects in environment has been presented but warrants careful monitoring
Coagulants (Fe^{3+}, Al^{3+})	1–30 ppm	Not used	Discoloration of receiving waters, burying of benthic organisms
Coagulant aids (polyacrylamides)	0.2–4 ppm	Not used	No known effects currently
Antiscalants (polyphosphates, phosphonates, polymaleic acids, polyacrylic acids)	1–2 ppm	1–2 ppm	Risk of eutrophication with polyphosphates
pH	6–7	6–7	
Antifoaming agents (e.g., propylene glycol)	Not used	0.1 ppm	Persistence in environment due to slow biodegradation—no known adverse effect currently
Heavy metals	Iron, chromium, nickel, molybdenum	Copper, nickel	Can partition to sediments and accumulate over longer term; potential for uncontrolled release from sediments by exogenous triggers
Cleaning chemicals	Alkaline (pH 11–12); acidic (pH 2–3); contains detergents, chelants, oxidants and biocides	Acidic (pH 2) solution containing corrosion inhibitors such as benzotriazole	Slug releases can be problematic; biocides could have impact on microbial communities with unknown consequences; chelants can mobilize metals from sediments

4.6.3 Brine Discharge from Brackish Water Plants

Brackish water desalination is usually carried out to reduce levels of chloride or nitrate. The salt concentration of the discharge is typically lower than that of seawater. Recovery is limited by the presence of scale forming compounds such as silica, or calcium. These discharges typically contain many of the same trace compounds, such as antiscalants and chlorinated by-products, as seawater RO discharges but higher amounts of organics, silica, phosphates and ions such as calcium and fluoride. Ion toxicity related to elevated levels of calcium and fluoride has been reported for brackish water concentrate discharges in Florida (Mickley 2000). Israeli experience has correlated disposal of brackish water concentrate in Ashkelon with elevated nitrogen and reduced oxygen levels in the receiving waters. It has also shown that brine discharge can disturb the ecology of the receiving waters. For instance, elevated silica in brine discharges seems to have encouraged selection for silica-accumulating algae in the proximity of the outfall.

4.6.4 Energy Use in Desalination

All desalination technologies utilize either thermal or electrical energy to separate water from salt. The environmental impacts of the energy use are dependent on the energy source. Currently, the sources are primarily fossil fuels based. Table 4.3 summarizes the energy use associated with the most common desalination technologies.

For a given fuel source or a mixture of such sources, air emissions associated with desalination will be lower for RO than for the thermal techniques. However, if a source of low-grade energy were available, such as excess steam that would normally be exhausted to the environment, to provide the thermal energy required, the differences in air emissions between RO and thermal methods would have to be analyzed on a case-by-case basis.

4.7 Mitigation of Environmental Impacts

The anticipated acceleration in growth of desalination has created an urgency to proactively address the associated environmental impacts. Detailed below are some strategies that have been advocated or adopted to mitigate those impacts.

4.7.1 Reducing the Impact of Seawater Intake Structures

Seawater intake structures are broadly of two types: open intake or subsurface. Open intake structures are comprised of intake pipes or channels and means to pre-screen

Table 4.3 Energy for desalination

Technology	Thermal energy use (MJ/m^3)	Electrical energy (kWh/m^3)	Comments	Total fuel energy (MJ/m^3)	Fuel (1 kg = 40 MJ/m^3)	Reference
Multiple stage flash distillation	200–380	3.6–4.4	The range in thermal energy reflects the energy use associated with either dedicated steam or use of low quality steam in a power plant	246–421	6.15–10.52	Afgan et al. (1999)
Multiple effect distillation	200–380	2.3		224–404	5.6–10.1	Afgan et al. (1999), Kennedy et al. (2002) (Electrical energy)
Seawater RO	None	3–6.5	Range encountered in operating plants	31–67	0.775–1.675	Kennedy et al. (2002)
Advanced seawater RO		2.0	Average power consumption at the most economic point	20.5	0.51	ADC (2008)

the water. These structures have the potential, if improperly sited or designed, to result in high impingement and entrainment of aquatic organisms (Fedorenko 1991). To guard against this, sites with low densities of marine organisms should be chosen. Areas such as estuaries and subtidal reefs with growth of macrophytes that serve as nurseries for other marine species should be avoided. Particular care should be taken to avoid spawning areas. Intake structures should also be located away from topographic features that promote turbulence and eddy formation as these may promote fish entrainment.

In an open intake system, the water has to be filtered prior to it being sent to the plant. This first level of filtration is commonly carried out using vertical screens. Vertical screens consist of wire mesh panels having openings of 6–9.5 mm. These are designed to rotate out of the sea periodically, every 8 h or so, to allow the accumulated debris to be washed. Fish pinned against these structures face high mortality rates.

Strategies for reducing such impingement include environmental modifications to discourage fish from approaching the screens too closely. Methods include modifications in lighting, generation of sound or electrical fields, and generating air bubbles around the screens. The success of these measures is species specific. Other alternatives include modifying the velocity field near the screens. Installation of velocity caps at intake pipe terminations and vertical louvers on screens in a direction perpendicular to fluid flow are some examples. The use of stationary screens with air backwash has also been reported.

Another innovation is the Ristroph screen, a traveling screen fitted out with baskets that allow fish and other organisms that stray too close to be safely removed. Ristroph screens have shown to improve impingement survival rates by 70–80% among various species (McLaren and Tuttle 2000).

Horizontally mounted passive screens, made of wedge wires having openings of 0.5–10 mm, have also been utilized with some success. They work best in environments where water movement occurs tangential to the screen. In all cases, the intake velocity of water is usually kept low enough to allow larger mobile organisms such as fish and turtles to swim away. Typically, this velocity is less than 0.15 m/s. Passive screens have demonstrated ability to reduce larval entrainment by 80% or more.

Subsurface structures are typified by the intake well being separated from the saltwater source by a natural or manmade permeable material. Examples of such structures are beach wells, collector wells, and infiltration galleries (Fig. 4.8). Beach wells are shallow intake wells that make use of beach sand as the filtration medium. Infiltration galleries are manmade structures of radial wells located in a trench and backfilled with a gravel pack and/or other selected filter materials. In some instances, pipes laid below the seabed floor acts as collectors. In general, subsurface structures are very effective at reducing both entrainment and impingement associated with water intake.

Other than the impact of intake structures on aquatic life, the construction process may also significantly impact the environment. The building of long pipelines into the sea or the burial of pipeline below the seafloor entails significant interac-

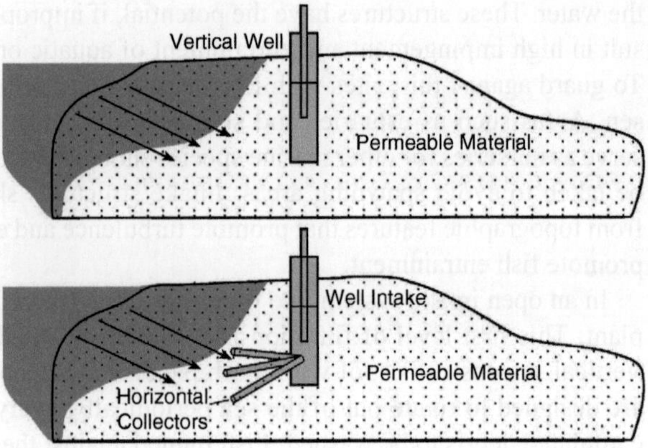

Fig. 4.8 Illustration of vertical beach wells and horizontal collector wells

tion with the existing environment and may be subject to complex regulations. One recent innovation that minimizes the impact of subsurface structures is the Neodren technology based on horizontal directional drilling. The below ground drilling process connects the intake wells with horizontal drains comprised of special porous filter pipes that are laid below the seabed. The seabed acts as the filter medium. The height of water in the intake well is maintained such that water flow is by gravitational head. The Neodren technology is reportedly in use at ten sites with a combined output of 300,000 m^3/day as of March 2007 (Peters and Pinto 2008).

In areas where existing power plants withdraw large quantities of seawater for cooling purposes, it may be beneficial to utilize the discharge to feed the RO plants. In this instance, there is no environmental impact due to construction of new intake structures or outfall structures. Additionally, there will be no incremental impact on aquatic life due to the RO plants. The Tampa Bay Plant in Florida, USA utilizes such a strategy as will the planned desalination facility at Carlsbad, California.

4.7.2 Reducing the Impact of Brine Discharge

The brine discharge from seawater RO plants is denser than seawater due to its higher salinity. In the absence of mixing, the higher density will cause the saline plume to sink to the bottom at the outfall location, affecting the aquatic life in that region. To avoid this, outfalls are designed to achieve rapid dilution of the brine in a large volume of seawater.

In RO plants collocated with power plants, the feed is usually a portion of the power plant cooling water discharge. In that case, the brine discharge is diluted with the much larger volume of cooling water discharged from the power plant (Voutchkov 2004). The combined discharge has a lower salinity than the RO discharge and is warmer. The combined effects of the higher temperature and relatively lower salinity allow the outfall discharge to mix with the entire water column with a less pronounced inclination to either float or sink. If the RO brine has a higher concentration of heavy metals due to corrosion of power plant condensers, specific pretreatment strategies to reduce heavy metals concentration prior to discharge or judicious choice of materials to minimize such corrosion may be required.

In some instances, particularly for small brine discharge volumes, injection into coastal wells may be an option. This option is not particularly attractive as these wells typically have limited storage capacity, and have life spans of about 10 years. Moreover, they have to be cleaned on a regular basis requiring a secondary back-up well when off-line. Other options that have been advanced to mitigate brine discharge impacts are brine evaporation and dilution with effluent outfalls.

4.7.3 Reducing Carbon Footprint of Desalination

There has been a concerted effort to reduce the carbon footprint of RO systems to mitigate global warming impacts. A two-pronged strategy to reduce the carbon footprint of desalination is currently underway. One is to reduce the energy consumption of RO, and the other is to use renewable energy.

4.7.3.1 Reducing Energy Consumption in RO

Energy consumption in RO is determined by two factors: (a) feed quality and (b) materials, equipment, and operational strategies. The influence of these factors on energy consumption and their management is discussed below.

Feed Quality

The main factor influencing energy consumption in desalination is the quality of the feed water. Feed water quality can vary significantly depending on location, current patterns, season, and pollution. Feed water characteristics affected are temperature, salinity, turbidity, organic content, and biological loading. Hence, the location and means of extracting seawater for desalination are two of the most important decisions undertaken in the design of desalination plants.

Extraction of seawater, as remarked earlier, can be carried out using either open intake structures or subsurface structures. Large seawater desalination plants invariably use open intake systems.

Beach well systems provide a much higher quality of water in comparison to open intake systems. Table 4.4 provides a side-by-side comparison of water extracted from a beach well and that obtained from an open seawater intake at Al-Birk, Saudi Arabia (Jamaluddin et al. 2005). The silt density index (SDI), a measure of

Table 4.4 Comparison of quality of open intake and beach well water at Al-Birk, Saudi Arabia

Water quality parameter	Beach well			Open intake
	#1	#2	#3	
As (ppb)	3.2	4.1	3.3	1.9
Fe (ppb)	14	12	15	20
Cu (ppb)	3.4	4.2	4.6	4.0
Pb (ppb)	9	12	12	13
Mn (ppb)	ND	ND	ND	0.5
Co (ppb)	ND	ND	ND	0.7
Zn (ppm)	ND	ND	ND	
NO_3 (ppm)	4	2	2	7
Hg (ppb)	ND	ND	ND	
Se (ppb)	ND	ND	ND	0.4
Cr (ppb)	ND	ND	ND	0.7
TDS (ppm)	44,029	45,653	44,375	40,967
pH	7.3	7.3	7.3	8.3
SDI	0.4	0.93	1.2	3.2
Turbidity (NTU)	–	–	–	2.1
TOC (ppm)	0.4	0.4	0.5	1

ND No data

the fouling potential of water, is lower in beach well water. This is due to a combination of lower concentrations of colloidal material and total organic carbon (TOC). The lower SDI allows operation of RO plants at higher flux rates between 9 and 10 gfd (15.3–17.0 LMH) compared to those operating on open seawater intakes (7–8 gfd). This allows a reduction in energy consumption. Table 4.5 lists some desalination plants drawing water from beach wells.

Not all locations are favorable for siting beach wells and well yields can vary significantly with location. Additionally, beach wells are limited by their yield potential. Yields of vertical beach well systems vary from 400 to 4,000 m^3/day and those of a Ranney type collector system between 8,000 and 20,000 m^3/day. Additional factors such as thickness of beach deposits, potential for beach erosion, and the potential for contaminating nearby fresh water sources are also important in determining the feasibility of beach wells for extracting seawater.

Table 4.5 Yields of some beach well intake systems used in desalination plants

Location	Intake volume (m^3/day)	Number of wells	Volume/well (m^3/day/well)
Bay of Palma, Spain	92,354	16	5,772
Ghar Lapsi, Malta	56,775	15	3,785
Ibiza, Spain	26,116	8	3,264
Telde, Spain	21,953	8	2,744
Morro Bay, CA	4,542	6	757

Choice of Pretreatment

Pretreatment of seawater is essential to minimize energy consumption. Improper treatment can lead to the deposition of colloidal materials, sparingly soluble salts, and biofilms reducing membrane permeability and sometimes product water quality. A 15% decrease in membrane permeability can increase the operational pressure by 6–10 bar (90–150 psi) thus increasing the energy needed for desalination.

Systems using open intake structures use treatment trains consisting of disinfection, coagulation/flocculation, and filtration processes. In spite of this treatment, there is still considerable risk of fouling and systems are operated conservatively. For example, flux rates for systems using water from open intakes are typically 7–8 gfd (11.9–13.6 LMH), lower than those using beach well systems.

Beach well systems reduce or eliminate the need for coagulation, filtration, and backwash cycles. This lowers both chemical and energy use and associated environmental impacts. It must be noted that not all feed water drawn from beach wells avoids the use of chemical pretreatment. In some instances, the presence of higher amounts of iron and manganese necessitates additional chemical treatment. In systems where iron and manganese are not a problem, water from beach well systems is typically chlorinated and filtered through cartridge filters. More recent practice is to avoid chlorination as this, paradoxically, enhances biofouling (Kammourie et al. 2008). While front-end chemical treatment is avoided, feed waters from beach well systems are typically low in dissolved oxygen and require aeration prior to discharge.

Large variations in feed water quality are best handled with pretreatment using microfiltration (MF) or ultrafiltration (UF) to produce high quality feed water for RO. Benefits ascribed to UF pretreatment are improved membrane life and reduced membrane cleaning (Pearce 2007). These benefits can outweigh the capital expenditures associated with MF/UF. From an environmental perspective, a physical treatment such as UF is preferable as it minimizes chemical use, handling, and disposal.

Managing Feed Salinity

The energy used in RO increases with the salinity of the seawater processed. In general, feed water salinity is not a variable that is easily changed. In some instances, a fortuitous reduction in feed salinity is obtained when water is extracted using beach wells (Ebrahim et al. 2001) though in other cases it can be higher due to aquifer geology. An interesting example of salinity management can be found in a proposed design for a desalination facility in Swansea, Massachusetts (Swanson et al. 2006). The plant proposes to withdraw feed water from the Palmer River during the 6-h period around low tide, and discharge concentrate during the 6-h period around high tide. The withdrawal around low tide results in lower salinity of feed water, achieving savings in energy. The discharge of brine during high tide minimizes the environmental impact by minimizing the salinity difference relative to ambient. A similar scheme is proposed for a planned desalination facility on the Thames in London, England.

Effect of Temperature

A reduction in temperature increases water viscosity and friction. The increased friction reduces the permeability of water through the membrane and increases axial pressure losses within the system. Under typical operating conditions of a seawater desalination plant, i.e., constant flux operation, an increase of temperature from 8 to 25°C can decrease specific energy consumption by 7.5–13%. Variations in temperature leads to both fluctuating energy needs and feed pre-treatment efficiency. In areas such as the Mediterranean, the temperature of seawater varies from 14°C in mid-January to 33°C in mid-August (Vial and Doussau 2003). In spite of the large effect of temperature on seawater desalination, it is not usually controlled.

Co-location with a power plant allows some flexibility in regulating the temperature within a narrower range and in using water warmer than the ambient. In contrast to seawater, water extracted using beach wells can show much lower variations in temperature. Reported data at the Ibiza desalination plant, for instance, show a near constant temperature of 19°C±1°C over a two-year period (Edlinger and Gomila 1996).

Devices for Energy Recovery

All major desalination plants currently under planning or of recent design incorporate energy recovery. The primary energy requirement in RO is for pressurization of water. Many intervening inefficiencies increase the thermodynamic energy requirement by a factor of 3–4. A major inefficiency in older designs was the loss of energy along with the brine. As an illustration, assuming the inlet pressure to an RO plant to be 60 bar, and pressure losses in the membrane and piping to be 2 bar, at 40% recovery, 60% of the flow emerges from the RO unit pressurized at 58 bar, representing 58% of the input energy. This energy, previously lost, is now recovered using energy recovery devices and results in reduced specific energy consumption in modern plants.

Energy recovery devices fall into two major categories: (a) centrifugal and (b) positive displacement (Stover 2006). Centrifugal devices in common use are the Pelton impulse turbine (PIT) and the hydraulic turbocharger (HTC) (Figs. 4.9 and 4.10). The PIT device is a collection of contoured buckets arranged around the edge of a wheel that is turned by the impact of a pressurized water jet. That rota-

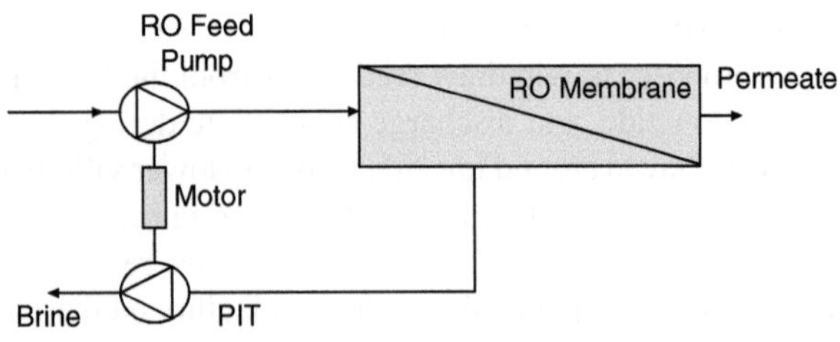

Fig. 4.9 A typical configuration of a Pelton impulse turbine in a desalination application

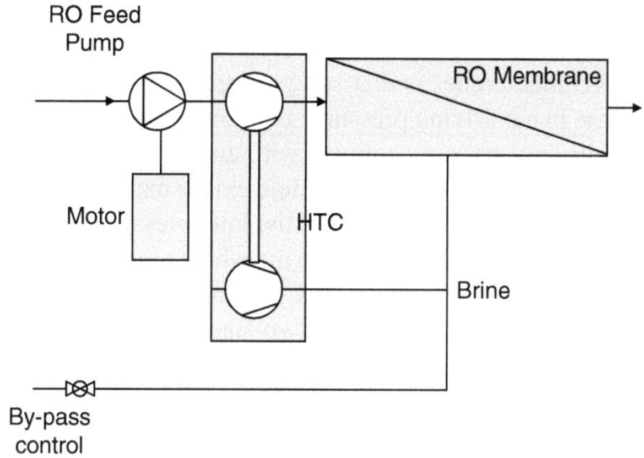

Fig. 4.10 Configuration of a hydraulic turbocharger in a desalination application

tional energy is transferred to the high pressure RO pump through an intervening driver. Pelton turbines have been used in a number of desalination plants and are considered a reliable technology. Turbine efficiency is relatively insensitive to inlet pressure variations and is of the order of 80%.

The HTC operates on a principle similar to the Pelton turbine. The turbine however is isolated from the feed pump and motor. The salt water, partially pressurized by the feed water pump, is further pressurized by the turbocharger. This allows much greater flexibility in operation and allows the feed pump to operate at its highest efficiency point.

Of the two positive displacement devices, one uses pistons mounted in equalizing pressure chambers to directly transfer energy from the brine stream to the feed stream. The feed pump pressurizes only the volumetric equivalent of permeate flow while the energy recovery device pressurizes the equivalent of the brine flow. The other positive displacement device achieves the transfer of energy without any pistons. The efficiencies of these devices are in the range of 90–95%.

Improving Motor/Pump Efficiency

The operational efficiency of the centrifugal pumps used to pressurize seawater is highest in a narrow range of specific speeds that is a function of rotational speed, volumetric flow, and developed head. The efficiencies increase with larger flows and lower heads. Given that, current trend is to use fewer larger pumps to feed multiple RO trains at a lower head. Pump efficiencies as high as 90% and motor efficiencies of 93% have been demonstrated in controlled studies.

Improvements in Operational Strategies and Materials

It has been noted previously that the net driving force in RO is the difference between the operational pressure and the osmotic pressure of the feed (see Fig. 4.3).

As the feed is dewatered, salinity increases. This increases the osmotic pressure and decreases net driving pressure for a set operational pressure. Another characteristic of RO membranes is that salt passage through the membrane increases with a decrease in net driving pressure. The combination of higher salinity leading to a lower net driving pressure, coupled with an increase in salt passage, necessitates the use of a higher pressure at the module exit to maintain both adequate permeate quantity and quality. This requires that the inlet pressure be raised as well. It has, therefore, long been a practice to operate the whole module under the higher pressure required at the module exit and use low flux high rejection membranes.

An alternative is to use a two-stage process with the concentrate from the first stage treated in a second stage (Veerapaneni et al. 2007). Using a combination of high flux, relatively low rejection seawater membranes in the first stage results in a greater quantity of permeate with low dissolved solids content at a lower pressure. The concentrate from stage 1 is further pressurized and dewatered to the final extent using a second stage. The membranes used in the second stage are tighter membranes with higher salt rejection. The volume of concentrate treated is lower than the feed by the amount of permeate produced in the first stage. The lower volume and the higher pressures required in the second stage allows the judicious application of energy only where required and achieves a reduction in specific energy consumption.

In the Long Beach method (Adham et al. 2003), a similar two-stage strategy is employed using nanofiltration membranes in both stages. However, it is the permeate from the first stage that is polished further in the second stage. The method is said to be capable of operating at 40% recovery with an overall salt rejection of 99%. The energy savings arise from the ability to significantly lower the operating pressure to produce a comparable quantity of permeate.

Finally, there is considerable ongoing research to increase the hydraulic permeability of the membranes. A doubling of permeability over 30 years of development has occurred resulting in reduced specific energy consumption (Birkett and Truby 2007). More recently, these efforts have been considerably boosted by use of novel materials that promise more rapid advances. One start-up company, NanoH20, claims a permeability increase of 100% through incorporation of nanocomposites into conventional polymeric membranes (http://www.nanoh2o.com). Work is also ongoing to fashion high permeability/high selectivity membranes from materials such as carbon nanotubes (Anonymous 2006b) and biological analogs such as aquaporins. Given these, the prospects for further reductions in specific energy consumption are bright and an energy consumption of <2 kWh/m^3 is likely to be the norm in the coming decade.

4.7.3.2 Renewable Energy

Another way to mitigate the environmental effects associated with energy use is to use renewable carbon or non-carbon energy sources where possible. These include biofuels, waste heat, solar, wind, tidal, and geothermal energy sources. Very few large-scale desalination plants have adopted this route but a few examples are emerging. An oft-cited example is the desalination plant in Perth, Australia, where a 130,000 m^3/day desalination plant is slated to supply about 17% of Perth's water. The specific energy consumption is expected to be 4.1 kWh/m^3 and will need 26 MW of power to operate the plant. The intention is to draw this power from the grid. The plant has advertised that it would offset this energy requirement with renewable energy purchased from the Emu Downs wind farm.

A similar scenario is emerging in Sydney, Australia, where 250,000 m^3/day of desalination capacity is being created. In the interest of cutting down on greenhouse gas emissions, the plant may only operate intermittently as a contingency measure in the event of drought. However, the plant operator has been directed by the New South Wales government to reduce greenhouse gas emissions by 100%. The plant is expected to do so by buying electricity from the Capital Wind Farm (http://www.sydneywater.com.au/WhoWeAre/MediaCentre/MediaView.cfm, accessed 24 November 2008).

While the above examples from Australia are similar, there are some interesting differences as well. In the case of Sydney, it is the desalination plant that is directly purchasing the electricity establishing a clear link between renewable energy use and desalinated water. In the Perth situation, the link is less clear as it is the power utility that is purchasing the power from Emu Downs as an offset. The probability of such purchase of renewable energy occurring regardless of the establishment of a desalination plant has left room for debate on whether the power required for desalination comes from a renewable source. Thames Water has committed to running its desalination plant (140,000 m^3/day) at Beckton, East London on 100% renewable energy, with biodiesel being one of the alternatives under consideration (Coffey 2008).

Clearly, considerable effort has been undertaken and is ongoing to reduce the environmental impact associated with desalination on a number of fronts. Many of the techniques and strategies that have been described previously have made large capacity desalination plants more eco-friendly and economical than ever before. One can anticipate that future developments will see further gains on these issues.

4.8 Membrane-Based Desalination at the Small and Medium Scale

In contrast to large capacity desalination plants, those that provide water to small communities in water-stressed and impoverished regions of the world face different types of challenges including unreliable energy supplies, shortage of skilled labor for plant operations and maintenance, lack of local availability of critical system components, and affordability. The feasibility of membrane based desalination technology in any given locale requires a careful analysis of the form and reliability of locally available energy, feed water quality, product water quality requirements, and local availability of appropriately skilled labor. A discussion on available energy sources, membrane processes, materials, and modules suitable for smaller communities follows.

4.8.1 Energy Sources and Their Suitability for Desalination

Figure 4.11 provides an overview of the most common energy sources and forms utilized for operating desalination equipment. Grid-based electricity is the preferred choice for operating RO plants, but it is frequently unavailable or unreliable in remote areas. Diesel is the most common fuel used for powering RO in remote locations. Among the listed renewable energy sources in Fig. 4.11, solar and wind powered RO are the most widely tested. The intermittent and variable nature of solar and wind energy poses additional challenges for operating desalination equipment.

4.8.1.1 Solar Thermal Desalination

Solar energy is typically harvested using either thermal collectors or photovoltaic cells that produce electricity. Flat plate thermal collectors are most common for low temperature (< 100°C) applications, and they are chosen for their modest cost, simplicity of construction, amenability to repair/reuse, and moderate efficiency. A salinity gradient pond is an alternative method to harvest solar energy. It can be used to drive both thermal and electricity based desalination systems where land is available and affordable.

Membrane distillation is currently the most promising small-scale solar thermal desalination technology. Demonstration membrane distillation plants producing 100–1,000 l/day of distillate have been operated successfully in Egypt, Jordan, and Morocco in the last few years (Rommel et al. 2007). A membrane distillation module, a flat plate solar collector, a photovoltaic module to power the pumps, valves and control system along with tanks for feed and product water constitute the main elements of the plant. In operation, salt water is pumped from the feed tank to the cold side of the membrane distillation module, where it condenses the water vapor that has migrated through the membrane from the feed side. In this process, the feed

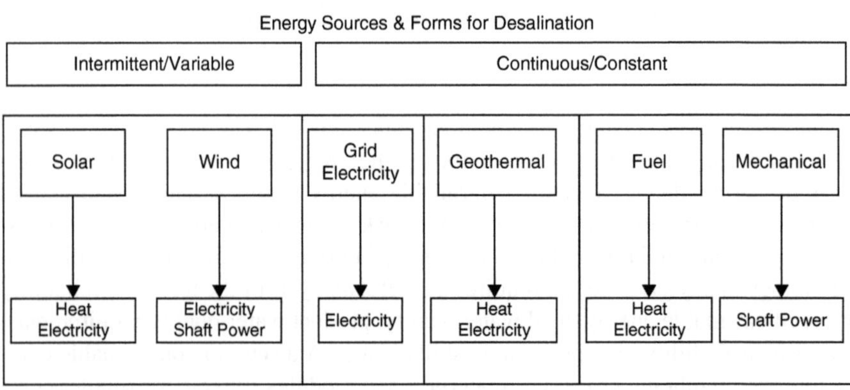

Fig. 4.11 Energy sources and forms potentially useful for operating desalination equipment

Table 4.6 Characteristics of a 100 l/day solar assisted membrane distillation system

Nominal capacity	100 l/day
Feed tank	500 l
Solar collector (single glazed collector/ no anti-reflective coating)	6 m^2
Header/risers	CuNi10 Fe
Photovoltaic module	80 W_{peak}
Membrane module area	7–10 m^2
Membrane material	Hydrophobic PTFE
Membrane length	450–650 mm
Module diameter	300–400 mm
Feed temperature in	80°C
Feed temperature out	25°C
Cooling water temperature in	20°C
Cooling water temperature out	75°C

water gets preheated via latent heat recovery. It is then directed to the collector, where it gets heated to the final process temperature and flows into the feed side of the module. The difference in water vapor pressure across the feed and condensate compartments provides the driving force for vapor transport. Direct contact between the distillate and cooling water is prevented through separation by a heat-transfer surface. The characteristics of a 100 l/day system are provided in Table 4.6.

One potential mode of failure in these systems is the possibility of water overheating in the solar collectors due to stagnation. This is partially addressed by locating the feed tank above the solar collectors ensuring the collector channels are always filled with water. However, this by itself does not preclude the possibility of steam generation and membrane failure by exposure to steam. To guard against this contingency, the water exiting the solar collector is isolated from the membrane module by a system that vents any steam formed. The same system also acts as a degasifier to purge air released from hot water during initial start-up. A second feature of these systems is the matching of the system capacity to the incident radiation. This is accomplished by modulation of the flow rate of the DC driven pump through a maximum power point converter. The control systems also ensure that system start-up occurs only after a minimum temperature of 50°C is achieved. A single pump feeds the membrane module and refills the feed tank, lowering system costs.

A larger system, with approximately 80 m^2 of solar collector area, nominally produces 1,000 l/day. It uses a modified design that separates the solar energy collection system from the saline feed water loop. Fresh water is used in the solar energy collection loop, and the required process heat is provided through heat exchange. The use of fresh water in the solar energy collection loop avoids the need for expensive corrosion resistant alloys.

One potential problem with membrane distillation modules is the inadvertent wetting of the membrane and loss of salt separation capability. This could occur when the membrane is exposed to water sources with low surface tension, such as those contaminated by surfactant. Hence, care is required to ensure such contaminants are not present or to provide some means for their removal.

A second concern with membrane distillation is the increased probability of mineral scaling due to the potentially higher water recovery and the higher operating temperature. Hence, robust control strategies and simple cleaning protocols are needed to ensure trouble-free system performance.

The future growth of membrane distillation looks promising because of its simplicity, dependence on heat rather than electricity, ability to handle a wide range of salinity, low pretreatment requirements, low pressure operation, and the elimination of feed sensitive RO membranes. Currently, the process is expensive, with a capital cost of $ 8,000 for a 100 l/day system and $ 52,000 for a 1,000 l/day system (Banat and Jawied 2008). The estimated cost of product water ranges from $ 30 to 36/m^3, but the final cost can be significantly reduced. For example, 40% of the estimated costs are related to monitoring and instrumentation and probably overestimate the cost of actual practices in the field. This estimate also discounts the use of existing local infrastructure to provide some of these required functions.

4.8.1.2 Solar Electricity Powered RO

RO systems have been successfully coupled with solar electricity in many demonstration plants (Garcia-Rodriguez 2003). A typical arrangement is shown in Fig. 4.12. The photovoltaic panels (PV) made of either crystalline or amorphous silicon produce direct current (DC) electricity. In some designs (Manolakos et al. 2008) this is used directly to power DC pumps. More commonly, the DC is converted to alternating current (AC) through use of an inverter to allow use of AC drives. DC drive pumps have been reported to be unreliable under continuous operation (Papapetrou et al. 2007). In addition, as solar energy is intermittent and variable, battery banks are usually employed to store sufficient energy to allow sustained RO operation and allow for smooth system shutdown. Deep discharge lead-acid batteries have been commonly used in most designs. In some designs, a backup energy source such as diesel has also been specified to allow system operation at night.

PV-RO systems of capacities ranging from 0.5 to 50 m^3/day have been designed or demonstrated. Capital costs of PV-RO systems are typically much higher than

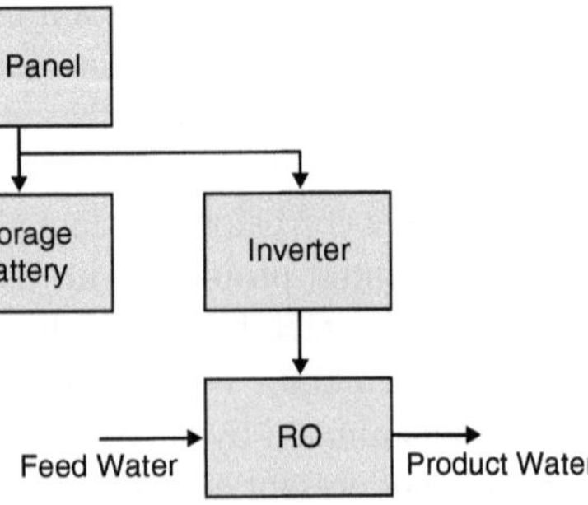

Fig. 4.12 Typical components of a PV-RO system

Table 4.7 Contributors to capital costs of a 3 m^3/day PV-RO system. (Tzen et al. 1998)

Component	% of capital costs
PV array	51.3
RO unit, tanks, piping, installation	19.5
Batteries	14.2
Inverter	3.9
Others (building, diesel back-up, etc.)	10.9

grid connected RO systems due to the additional PV arrays, batteries, and battery charging systems (Table 4.7).

Estimated costs of water production by PV-RO vary significantly based on location, feed water salinity, and plant capacity but in general is more expensive than grid connected RO or water transported over short distances. These high costs have deterred large-scale adoption of these systems.

Several approaches have been investigated to minimize these costs. One approach is to minimize the size of the PV array by minimizing the RO system energy requirement. Clark pumps have been reported to be suitable for energy recovery in small-scale systems. In one study of desalination of seawater, the incorporation of a Clark pump in a PV-RO system producing about 4 m^3/day of water resulted in an energy consumption of only 3.51 kWh/m^3; comparable to those achieved at much larger capacity plants (Thomson et al. 2001). The second approach lowers costs by eliminating the battery storage systems. Eliminating battery storage negates the need for charge controllers and battery monitors as well. It also eliminates the costs of battery replacement, a significant contributor to the operating costs of the system. Eliminating the battery storage requires that either of two approaches be taken: (a) specifying an RO system (and therefore a PV array) with excess capacity (b) modulating the flow through the RO system to match the available energy. The second approach is preferable as it allows capital costs to be reduced. Several studies have now demonstrated the feasibility of batteryless PV-RO systems designs incorporating the second approach for short periods of time (Manolakos et al. 2008; Thomson et al. 2001; ITN Energy Systems Inc 2004). These advances along with the decreasing cost of PV arrays (Helal et al. 2008) should help make PV-RO systems more affordable in the future.

4.8.1.3 Solar Electricity-Electrodialysis

The operation of electrodialysis systems with solar-generated DC electricity has been reported for treating brackish water (Adiga et al. 1987; Ortiz et al. 2007). The generated current, after power conditioning, can be used directly to run the electrodialysis system. Compared to RO systems, electrodialysis systems tolerate intermittent operation more readily, operate at lower pressure, and require less pretreatment. These characteristics make it a relatively robust operation and amenable to operation without battery backup. Desalting of seawater is usually not practiced due to higher energy requirements.

4.8.1.4 Wind Electricity-RO

The challenges associated with coupling of wind based energy with RO systems are similar to those faced with solar. It is therefore not surprising that similar solutions, such as power management, variable load operation, and energy storage, have been offered. Wind energy has matured rapidly in recent years and costs have fallen dramatically, particularly for larger scale projects. Though wind appears to be particularly suitable for supplying autonomous grids that serve larger geographical regions, wind as a stand-alone RO power supply, is more problematic for smaller scale systems, unless backed up by other energy sources. Wind-diesel hybrid systems appear to have potential in this regard. One Danish firm, Vestesen A/S, offers commercial wind-diesel RO systems ranging from 100 to 10,000 kW capable of powering desalination modules producing 1,000–20,000 m^3/day.

4.8.1.5 Wind Mechanical-RO

In these configurations, energy extracted from the wind is either hydraulically or mechanically coupled to the RO unit. Robinson et al. (1992) built a prototype system wherein the windmill directly operated a positive displacement piston pump. The piston pump pumped water to a series of pressurized hydraulic accumulators that acted as energy buffers analogous to batteries in an electrical system. The RO unit was operated when the hydraulic accumulators were able to deliver a specified flow rate of water at predetermined pressures for a minimum time period. Auxiliary power needs were met with a small PV panel. The membranes used in these trials were operated at pressures between 90 and 165 psi on brackish waters of 2,000–6,000 mg/l total dissolved solids (TDS). Water recovery was limited to 6–11% to minimize membrane fouling and pretreatment. In general, the mechanical components of the system appear to have operated fairly well with few problems. Production was reported to have deteriorated by 25% over the operational period of 13 months.

In a more recent study carried out in the Dutch Antilles (Rabionovitch 2008), seawater was desalinated using energy derived from a windmill mechanically coupled to a piston pump. The direct coupling avoids energy losses incurred in the back and forth of conversion of mechanical energy to electrical and back. The plant was designed to produce 5 m^3/day of desalinated water at an average wind speed of 7 m/s. The RO plant was designed to operate at a constant water recovery of 20% but with varying pressure and flow rates. The design also incorporated an energy recovery device that further lowered specific energy consumption to 5.2 kWh/m^3. The membranes were flushed with desalinated water prior to shut down to prevent membrane damage. Long term operational reliability information on this system remains to be collected.

4.8.1.6 Diesel-RO

Diesel operated RO systems are fairly common in military and disaster relief operations. However, diesel can be expensive and is polluting. Adapting diesel sets to run on locally produced energy such as biodiesel and biogas can go a long way to make desalination accessible in remote areas. The higher maintenance cost of diesel generators is one disadvantage that has to be overcome to ensure successful implementation.

4.8.1.7 Manually/Animal Operated RO

Manually operated RO systems are an interesting approach to completely cut the cord on energy dependence from external sources. The primary limitation is the ability to produce large quantities of water. However, production of drinking water from both brackish water and seawater is practical. Manually operated RO systems are sold in niche markets for survival at sea or for people traveling through remote areas.

Two patents (Manjikian 1978; Bray 1978) describe the use of manually operated RO devices. One (Manjikian 1978) provides the means to convert the rotary motion of a cranking device to a reciprocating motion to pressurize a feed stream. A separate mechanical link utilizes the rotary motion to rotate the membrane as well. Rotation of the membrane surface helps reduce the build-up of high concentrations of salt at the membrane surface and increase the quality of water produced at a given operational pressure. A prototype based on the patent, shorter than 3 ft in length and less than 30 pounds, was able to produce 1 l/h from seawater. The second patent utilizes a hand-operated lever to generate a reciprocating motion to both pressurize the feed water and move the membrane relative to the housing to reduce the build-up of salt. No prototype appears to have been built. Compared to the second design, the first design provides a larger amount of membrane surface area for a given housing length.

A more recent patent (Pathak et al. 2008) envisions the RO unit to be run using animal power. The high torque, low speed power produced by a pair of bulls moving in a circular path is utilized to power a high pressure triplex pump, producing 20 l/min at 300 psi. When connected to an RO unit, brackish water of 5,000 mg/l TDS was converted to treated water of 600 mg/l TDS at 3 l/min. The design utilized did not recover energy. Water production was reportedly sustained for 4 h. In summary, optimization of manually/animal operated systems for affordability, efficiency, and robustness holds much promise in making desalination technology much more accessible.

4.8.2 Membrane Material Choices

RO membranes are the most critical component in any desalination plant. Unfortunately, they are also the most vulnerable to mishandling, and susceptible to mechanical and chemical damage. In highly cost-sensitive situations, membrane replacement costs alone can make membrane-based desalination unaffordable. It is in this context that there arises the need for materials that lend themselves to being locally manufactured into membranes. There is good reason to believe that such materials can form the basis of a desalination cottage industry.

Cellulosic derivatives such as acetate esters were the first to be successfully used for membrane desalination. Cellulose is the main component of higher plant cell walls and one of the most abundant materials on earth. Typical sources of cellulose include wood pulp, cotton fibers and cotton linters. Cellulose can also be obtained from various plant fibers, such as corncobs or stalks, soybean hulls, sugar cane bagasse, oat hulls, rice hulls, wheat straw, sugar beet pulp, bamboo, and fibers such as jute, flax, and ramie among others (Hanna et al. 2001). Bacterial sources of cellulose have also been developed using *Acetobacter xylinum* that ferment substrates of glucose from corn syrup (Son et al. 2001).

If local manufacture of cellulosic membranes is envisioned, it may be based on either a cellulose rich substance or cellulose acetate. Cotton and waste paper are examples of materials that can be readily converted to cellulose acetate. The three main steps in the manufacture of cellulose acetate are: (a) swelling of cellulose (b) esterification, and (c) partial hydrolysis (Fig. 4.13).

The cellulose acetate can then be processed into membranes with an asymmetric structure with relative ease. The normal route is by dissolving the cellulose acetate

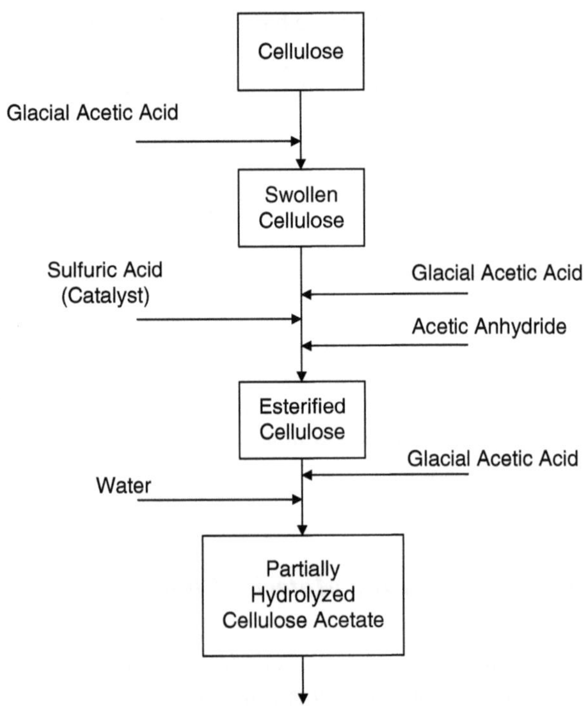

Fig. 4.13 Steps in the conversion of cellulose to cellulose acetate

in a suitable solvent, in combination with a pore-forming compound. The solution is then cast to form a film. The solvent is removed through a combination of evaporation and leaching, under conditions that allow the polymer to coalesce and form a specific structure. For desalination, a thin, dense, top layer and a porous substructure are desired for effective water transport and low salt passage. By varying the recipe and operational parameters, i.e., temperature, humidity, and post treatment strategies, the processor can optimize desired functional properties such as porosity, water transport, and salt rejection. The science of structure formation may be complicated, but the art of manufacturing is well known.

Another material that satisfies the requirement of local availability and utility for desalination is chitosan. Chitosan is a biopolymer of N-acetylyglucosamine, and it can be produced from chitin, which is the main component of the exoskeletons of crustaceans such as crabs, lobsters, and shrimps. Large amounts of these materials are frequently disposed of as waste. It is estimated that 60–80,000 tons of chitin may be annually available from shrimp production in India alone (Shetty et al. 2006). Chitosan is the deacetylated form of chitin, and it is more easily cast into films than chitin. Yang and Zall (1984) cast RO membranes from chitosan by first dissolving it in a 2% acetic acid solution, and casting it on a glass plate to form a clear chitosan salt membrane. The membrane was then neutralized with 10% sodium hydroxide. Such membranes are insoluble in alkali solutions in concentration up to pH 13 but still fairly soluble in dilute acids. Acetylation of the free amino groups in chitosan provides acid stability. To achieve this, the membranes prepared previously were acetylated with a solution containing five parts of acetic acid, three parts of dicyclohexylcarbodiimide and 100 parts methanol. These membranes exhibited salt rejections ranging from about 80% at sodium chloride concentrations below 1% to about 25% at concentrations of 3% at pressures ranging from 300 to 500 psig. Salt rejection was insensitive to pressure at all but the highest salt concentration and even then only slightly. Flux rates were of the order of 5–7 LMH at pressure of 300 psig for salt concentrations of 3%. While the observed membrane flux and salt rejection properties would be considered poor relative to current commercially available membranes, they would be useful for brackish water desalination. Moreover, it is highly probable these properties can be improved through further optimization to an extent sufficient to enable a two-stage process for desalination of seawater.

Alginic acid, a product of seaweed, has been used as a coating material for membranes used in membrane distillation, as well as in combination with chitosan for pervaporation and RO membranes. In one study, a 6% aqueous sodium alginate membrane was cast onto a glass plate and dried. Following this, it was crosslinked through immersion in a copper nitrate solution to form a copper-alginate membrane complex. The resulting membrane exhibited a flux of approximately 3 LMH and a rejection of about 78% at 882 psi when used for desalinating a sodium chloride solution of 0.05 M (0.3%) (Kim et al. 1987).

The possibility of using natural minerals and clay-type material has also been explored. Clays are known to exhibit salt rejection properties. For example, 0.5 mm thick montmorillonite membranes were reported to have sodium chloride rejections as high as 90.3% at salt concentrations of 1 mM (Ishiguro et al. 1995). This rejection decreases to 50% when salt concentration increases to 100 mM with a corresponding decrease in permeability from 0.4 LMH to about 0.35 LMH at 450 psi. Others have investigated the utility of bentonite clay for desalination (Liangxong et al. 2003; Whitworth et al. 2003). The results to date suggest that natural clay materials behave as charged membranes—i.e., allow greater salt passage when exposed to higher salt concentrations—and have low permeability. These characteristics limit the potential of natural clay materials on their own for desalination. However, they may be useful in dynamic membranes.

Another pathway to making membranes locally is to form them "dynamically." In this approach, a porous support is infused with an inorganic or organic colloidal solution. The colloidal solution partially occludes the pores, allowing it to function as a semi-permeable barrier. Membranes produced in this manner span the application range from microfiltration to RO. The primary advantage of these membranes is their ability to be formed in-place and regenerated as desired, allowing the user flexibility in maintenance.

Many materials, such as ferric hydroxide and other inorganic hydrous oxides, bentonite, humic acid, finely ground ion-exchange resins, poly acrylic acid and polyvinyl pyridine, have been shown to form dynamic membranes with varying salt rejection properties and permeability (Kraus et al. 1967). The mechanism of salt rejection exhibited by the above materials is charge-based, i.e., salt rejection decreases with an increase in salt concentration in the feed. For that reason, dynamic membranes may be more suitable for brackish water desalination or for softening. However, there is no reason to conclude that dynamic seawater desalination membranes cannot be prepared.

In summary, past research indicates the potential of materials such as naturally derived polymers, and inorganic materials to function as desalination membranes. However, extensive development work will be required to translate that knowledge into effective materials that can be fabricated in a simple, ecologically efficient manner at a local level.

4.8.3 Membrane Modules

Membranes are typically packaged into modules in various configurations. These configurations significantly influence capital and operational costs. Currently, the spiral wound configuration is widely used in desalination applications. It offers multiple advantages such as high packing density, low pressure drop, and low initial costs. Moreover, a wide array of accessories for spiral-wound membranes from multiple suppliers exists, ensuring a stable low-cost supply chain.

However, the spiral wound membrane is susceptible to having its flow channels obstructed by fine particulate material, biofouling, and mineral scale. To guard against these, feed water pretreatment is usually required as well as close monitoring. When a module gets plugged, restoring it to operational status requires skill.

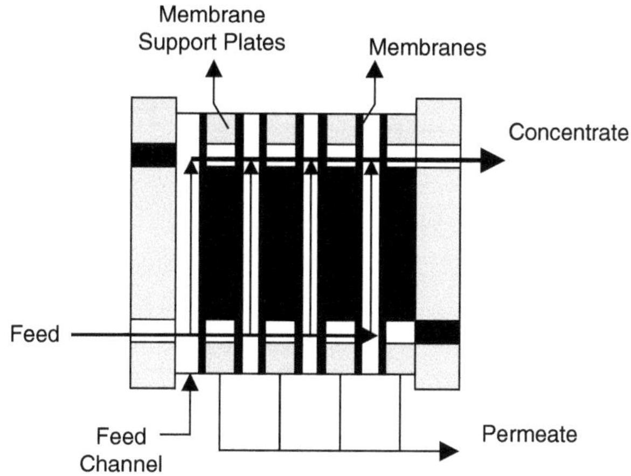

Fig. 4.14 Schematic of a plate and frame membrane module

To make these modules more useful in cost-sensitive situations equipment that permits the module to be disassembled and reassembled multiple times to facilitate cleaning would be needed. To achieve this would require a winding device that can form the spiral in a consistent manner as well as a means to reliably seal the membranes on a repeated basis. This would require additional development work.

Another configuration, the plate and frame type, offers access to membranes and modularity. Its disadvantages are low packing density and unreliable sealing under high-pressure conditions. However, the low packing density is not a major deterrent in small scale applications. Reliable sealing has been demonstrated in commercial units manufactured by the DDS company of Denmark. Figure 4.14 is an illustration of a plate and frame unit. The membranes are supported on both sides of a plate that also act as a permeate collector. The feed flows through the channels formed between the membrane support plates. The plates in turn are assembled in a frame to form a stack. These modules can be disassembled for cleaning and reuse, and they can be made from a wide variety of materials such as wood, metal, and plastic depending on the operating pressure. Only membranes need be replaced, if necessary.

There are also opportunities for unconventional designs, especially for systems that serve individuals. For example, one can conceptualize a multiple stage RO process based on inexpensive construction materials such as soda bottles and bicycle pumps (Fig. 4.15). The soda bottle can serve as both a feed reservoir and a pressure vessel. The cap can be modified to hold loose RO or NF hollow-fiber membranes with an external separation layer. Partially desalinated water can be produced by pressurizing the system with a bicycle pump to modest pressures of the order of 100 psi. If necessary, the water can be reprocessed to attain potable water quality.

A second approach might be to use a forward osmosis type system that avoids pressurization. Trials in Uzbekistan (Khaydarov and Khaydarov 2007) have demonstrated the successful use of diethylether as the draw solution for desalinization

Fig. 4.15 Schematic of a potential low cost RO system

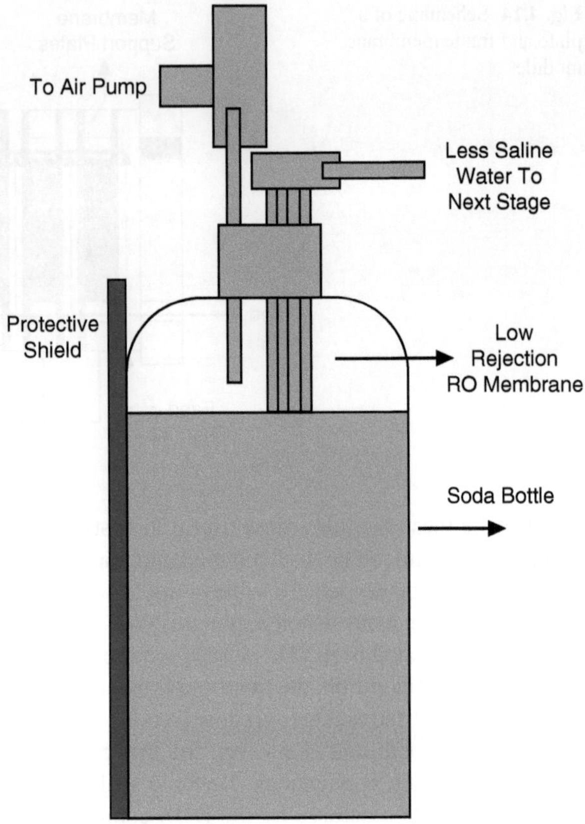

of brackish water in a forward osmosis mode (see Fig. 4.7). The primary drawback is the lack of chemical resistance of the TFC membranes used and thus the need for frequent replacement. However, if the membranes are made chemical resistant or inexpensively, the process becomes more feasible.

4.9 Integrated Approaches

In order to meet the Millenium Development goal of supplying safe drinking water, a comprehensive approach that recognizes the linkages between energy, water, and environmental protection will be necessary. Moreover, such approaches must address issues related to access to capital resources and creation of economic opportunities. Figure 4.16 illustrates one such approach (Hodges et al. 1993). It highlights the integration of aquaculture and halophyte irrigation to produce food, fiber, oil, water and power in an environmentally responsible manner.

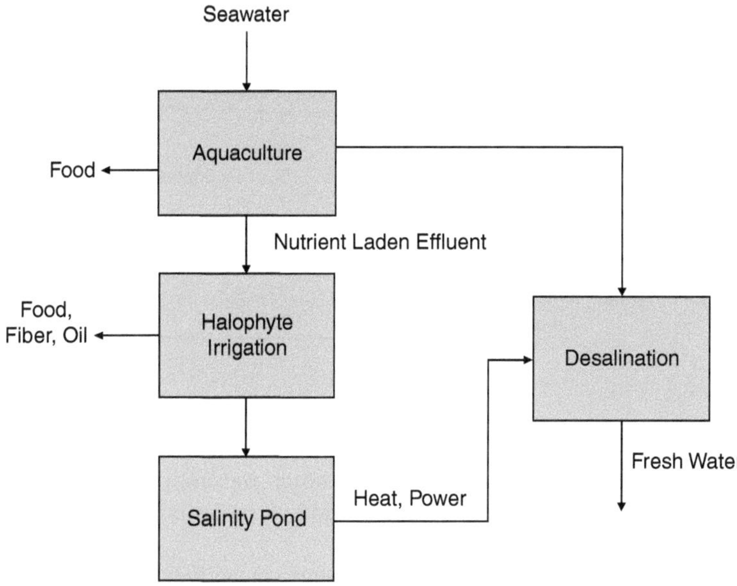

Fig. 4.16 An integrated scheme for a coastal desert incorporating desalination

The use of seawater for shrimp production is a widespread practice in many countries that generates jobs and income. Unfortunately, the practice also results in a nutrient rich effluent that if directly discharged to the ocean causes pollution. An alternative is to utilize this nutrient rich effluent to irrigate halophytes for food, fiber and oil production.

Halophytes are plants that tolerate and even thrive in highly saline waters. It is estimated that about 3,600 species of halophytic plants exist throughout the world (Yensen 2006). Salicornia in particular has received attention since the 1980s, due to its potential use as an oilseed crop. Breeding programs at the University of Arizona have resulted in varieties that have crop and oil yields comparable to conventional cultivation of soybean. The plant yields an estimated 851.7–946.4 l (225–250 gal) of oil/ha (Christiansen 2008). Oil quality and composition are comparable to that of safflower (Table 4.8). A portion of the seed oil can be converted to biodiesel, to provide emergency fuel for desalination systems that are powered by intermittent sources such as PV or wind. After extraction, the remaining seed meal can be used for animal feed with additional processing (Wu and Sessa 2004). The biomass is also claimed to be suitable for use as fuel or as soil amendment after composting.

The effluent after halophyte irrigation is nutrient depleted and can be routed to an evaporation pond for subsequent use in a salinity gradient pond for collecting solar energy. These ponds can provide both electricity and heat, which in turn can power desalination systems. This concept was put into practice in the Seawater

Table 4.8 Composition of Salicornia seeds

Composition	% of seed weight
Oil	26–33
Protein	30–33
Fiber	5–7
Ash	5–7
Fatty Acids	% of fat content
Palmitic	7.7–8.7
Stearic	1.6–2.4
Oleic	12–13.3
Linoleic	73–75.2
Linolenic	2.4–2.7

Farms Initiative in Eritrea (Hodges 2004), and it appears to have been successful, before encountering difficulties due to war. Desalination was not part of the project, though it would have been relatively easy to add on by utilizing the existing salt pond.

A second example illustrating the use of economic incentives to provide access to clean water can be found in the Maldive Islands. Solco, Ltd., an Australian company, and Triodos Bank Renewable Energy Development Fund, Netherlands, have operated a PV-RO system on the Island of Kulhudhuffushi capable of producing 1.5 m^3/day of water. The project was conceived to be financially self-sustaining through sale of water. Estimated project costs are shown in Table 4.9. The payback is estimated to be approximately 3.5 years. The project meets a well-defined need of providing clean water in an island community in an ecologically conscious manner while generating local employment. In this case, the produced water offered an alternative to imported bottled water. This market-based approach provides an alternative model to making water available, if not exactly affordable.

Table 4.9 Estimated project costs of PV-RO plant in Maldives

	US$	Comments
Capital costs	$ 130,000	Capacity—1.5 m^3/day product water
Principal and interest	$ 35/day	
Average maintenance	$ 11/day	
Site rental	$ 1.70/day	
Daily operational costs	$ 15/day	
Daily labor costs	$ 10/day	
Total costs	$ 72.7/day	
Product sales	$ 187.50	$ 125/m^3
Cash flow	$ 115/day	

4.10 Conclusions

The accelerating demand for water calls for a major global effort to harness, conserve, and add to existing water supplies. Membrane based desalination is expected to play a major role in augmenting and reusing existing water supplies. However, this will also require the development of scientifically sound strategies to mitigate the environmental impacts associated with desalination plants. Multiple means have been developed to mitigate the adverse impacts of desalination, especially in the area of energy consumption. Advances in membranes, energy recovery devices, and operational strategies have contributed to reductions in specific energy use. Nevertheless, the total energy consumption will grow dramatically as desalination capacity increases. The use of renewable carbon-based energy or non-carbon energy, such as solar and wind, to power desalination plants will therefore be essential to minimize greenhouse gas emissions and associated climate change effects.

While the prognosis for environmentally conscious desalination is good for large-scale plants, it is not clear whether these advances will materially contribute to the welfare of the most disadvantaged segments of society. If the past is any indication, we think not. We suggest that the most effective approach to provide clean drinking water in these communities is to empower individuals and small communities to provide for their own safe water supply. Membrane based desalination is one tool among many that can contribute to that empowerment. Even with appropriate technology, the chances of success are greatest when desalination is integrated within a larger effort that addresses key economic, social, and environmental issues.

Where membrane based desalination has a role, it is best implemented by coupling locally available energy resources with robust systems and materials that enhance local autonomy and self-reliance. It should be specifically recognized that the benchmarks currently in place for assessing membrane efficiency in a large-scale setting are not applicable to the small scale. The use of esoteric materials and complex manufacturing should be exchanged for easily accessible materials and cookbook recipes that are executable with rudimentary equipment. Future work needs to concentrate on converting a variety of locally available materials and processing them into useable RO and ultrafiltration membranes in a simple, safe, and cost effective manner. Operations that require conditions far from ambient in terms of temperature and pressure should be avoided by embracing the idea of multiple processing and fit-for-purpose quality benchmarks. Finally, the experience gained in developing appropriate technologies in the fields of construction, sanitation, and irrigation should be relied upon to form the bedrock for a similar effort aimed at decentralized desalination.

Chapter 5
Bank Filtration as Natural Filtration

Chittaranjan Ray, Jay Jasperse and Thomas Grischek

Abstract When wells are placed close to a surface water source (such as a lake or a river) and pumped, a portion of the surface water is induced to flow to the well. As the water travels from the river to the well through the riverbed sediments and underlying aquifer material, suspended and dissolved contaminants of surface water are "naturally" filtered out using a combination of physical, chemical, and biological processes. If the surface water is a river, the system is called riverbank filtration (RBF). If a lake serves as source water, the system is a lakebank filtration. These natural filtration systems have been operating for more than 100 years in Europe and for over half a century in the United States, providing safe drinking water to communities. For the RBF systems to work effectively, there must be a hydraulic connection between the river and the alluvial aquifer where the wells are located. Unclogged river bottoms are ideal for RBF operations. RBF systems are known to remove turbidity, microbes, and chemicals present in surface water and the removal efficiency is a function of well location, pumping rate, source water quality, etc. A fraction of dissolved organic carbon (DOC) is also removed which helps in reducing the formation potential of disinfection byproducts during chlorination of the filtrate from RBF systems. RBF systems can be adapted to a given site using engineering judgment. Use of inflatable dams to raise water levels in rivers in low-flow periods can augment well yields. Similarly, diverting a part of the water from the river to an infiltration basin and strategically placing wells between the river and the infiltration basin can enhance yield. Over the years, several improvements to the design and construction of the RBF systems have taken place. Use of these methods at future sites can improve the efficiency of RBF. Besides siting issues, periodic maintenance and early-warning systems to monitor river water quality are needed for sustainable operation of RBF systems. RBF has one of the best potentials to be used as a natural filtration system in populated riparian communities in developing countries. This chapter presents some of the advantages and

C. Ray (✉)
Civil & Environmental Engineering, University of Hawaii at Manoa,
2540 Dole Street, Holmes Hall 383, Honolulu, HI 96822, USA
e-mail: cray@hawaii.edu

C. Ray, R. Jain (eds.), *Drinking Water Treatment*, Strategies for Sustainability,
DOI 10.1007/978-94-007-1104-4_5, © Springer Science+Business Media B.V. 2011

limitations of using RBF for water treatment. Design, cost, maintenance, and future research needs are presented.

Keywords Aquifer storage and recovery • Bank filtration • Collector well • Hyporheic zone • Pharmaceutical compounds • Microscopic particulate analysis

5.1 Introduction

When the source of drinking water for a community is surface water (e.g., river, lake, or reservoir water), the water is usually treated in conventional purification facilities to remove suspended and many dissolved contaminants. The cost of treatment and the quality of treated water depends on the quality of source water, the treatment unit operations employed, and the cost of energy used in treatment. For developing economies, construction and maintenance of large treatment facilities is not only capital intensive, but the successful operation of these facilities requiring trained operators is also a challenge.

Groundwater, when used for drinking, often does not need treatment beyond disinfection unless it is excessively "hard" (having high concentrations of calcium and/or magnesium ions present as sulfates, carbonates, and/or bicarbonates), contains dissolved iron and manganese, or is contaminated with anthropogenic and/or natural chemicals. However, the availability of groundwater is a problem for communities where the aquifers are marginal (e.g., hardrock areas).

So called *bank filtration systems* (see Chap. 2) offer a cost efficient mechanism for improving surface water quality. For a bank filtration system to function properly, a good quality source water must be present and there must be a hydraulic connection between the river/lake and the surrounding alluvial aquifer. When wells are placed sufficiently close to a surface water source and pumped for a period of time, a portion of the surface water is induced to flow to the wells. During soil and aquifer passage of the induced water, many suspended and dissolved contaminants present in the surface water are removed. The bed sediments and underlying soil and aquifer material act as natural filters for removing various contaminants from surface water. A schematic diagram of riverbank filtration (RBF) is presented in Fig. 5.1. Although the following discussions will focus primarily on rivers as the source water for these RBF systems, the underlying principles of "bank filtration" remain constant regarding the use of any adequate surface-water supply, e.g., lake, infiltration basin, etc.

As shown on the plan view of this figure, wells are present in an aquifer adjoining a river. Utilities use either horizontal collector wells, vertical wells, or a combination of both depending on site hydrogeology and pumping needs. Monitoring wells may be present on the landside to examine the quality of groundwater entering the bank filtration systems. The riverside monitoring wells are used to examine the quality of the water that infiltrates from the riverside and its transformation process as the infiltrated water moves to the wells. In the cross-sectional view of this figure, a horizontal collector well is shown with several laterals and a central caisson to

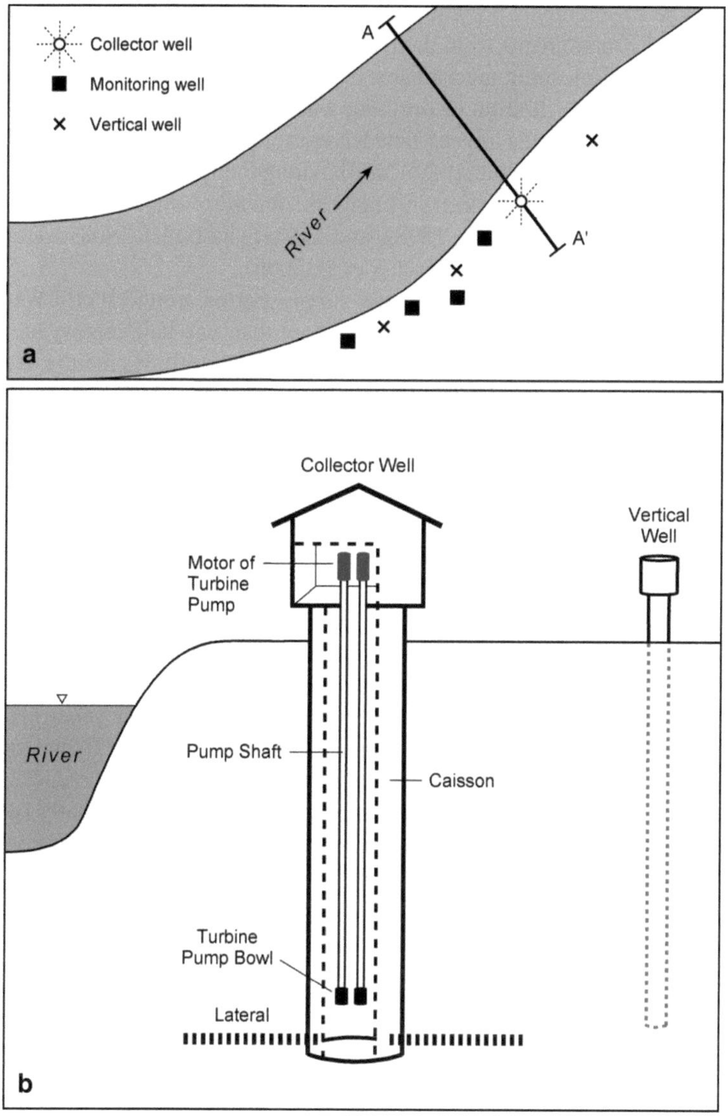

Fig. 5.1 Schematic of a riverbank filtration showing production and monitoring wells **a** Plan view, **b** A-A' cross-section

which the laterals are connected. One or more high-capacity pumps are installed directly above the caisson to pump the water out.

For an RBF system to function properly, the river should be hydraulically connected to the aquifer. This ensures infiltration of river water to the aquifer under pumping stresses from the wells placed in the adjacent alluvial aquifer. River hydraulics and morphology of the deposited materials determine the bed load movement and the deposition pattern of sediments on the riverbed. These factors affect the amount and type of sediments present on the riverbed; which ultimately affect the flow of water and contaminants from the river to the pumping well. The annual floods in rivers act as self-cleaning mechanisms to move the deposited sediments.

The RBF form of natural filtration of drinking water has been practiced in Europe for more than a century. In the lower Rhine River region, RBF for water supply has been operating since 1870 (Schubert 2002a, b). Along the Elbe River, there are three other RBF systems that have been in operation for more than a century—Saloppe (since 1875), Tolkewitz (since 1898), and Hosterwitz (since 1908) water works in the city of Dresden, Germany (Fischer et al. 2006).

In the United States, RBF for industrial water supply started around World War II. RBF has been used for public water supply for more than one-half century at a number of locations in the United States. While European water utilities mainly use vertical wells, most utilities in the United States that have installed RBF in the past four or five decades use horizontal collector wells (Ray et al. 2002b).

Natural filtration has become a significant part of the water supply for many cities in Europe and the United States. In Germany, a number of large cities located along the Rhine and Elbe rivers extract drinking water through RBF. The Slovak Republic and Hungary rely heavily on natural filtration of the Danube River water for supplying their major cities. Cities in Slovenia, Switzerland, and France also rely on natural filtration for their water supplies. A growing number of cities in India and South Korea are presently resorting to RBF for their water supply.

One of the main advantages of natural filtration employing RBF includes possible reduction of water treatment with the exception of chlorination or other disinfection. Even if the RBF water is further treated at an existing water treatment plant, the amount of chemicals needed for coagulation and flocculation is reduced. Compared to surface water, the bank filtrate temperature fluctuation is greatly reduced. This facilitates the control of coagulation reactions and lengthens filter runs. Elimination of a conventional filtration plant can reduce significant capital investment and associated operation and maintenance costs. RBF also enables the water systems to deal with chemical spills in source waters as the surface water is not directly drawn to the treatment units.

Natural filtration can be used as a pre-treatment for reverse osmosis membranes. In doing so, a large portion of the organics and particulate matter are removed during natural filtration prior to loading the water to the reverse osmosis units. This pre-treatment can reduce membrane fouling. Natural filtration can also be useful for producing high quality water for irrigated agriculture, aquaculture, fish hatcheries, and for recreation along polluted stretches of rivers.

Because RBF systems are more directly affected by natural conditions (e.g., droughts or floods) than conventional water treatment systems, careful planning and operational monitoring are required to assess natural variability of riverbed and river water quality. Uncertainty exists about the quantity of water and the quality of water to be produced from an RBF site until the system is built and operated. Scouring and clogging of the riverbed can affect the water quality after the system is built. Thus, careful study is needed to determine a place for siting RBF systems where such problems are less likely to occur.

5.2 Natural Filtration's Implications for Sustainability

When compared to conventional purification systems, RBF systems offer multiple advantages in terms of sustainability because RBF systems utilize natural filtration. These advantages include lower energy and resource requirements, little or no generation of waste streams, reduced environmental impacts during construction and system operation, and increased adaptability to changing water supply conditions due to climate change.

5.2.1 Energy Consumption, Resource Requirements and Waste Generation

Whether RBF is used as pre-treatment or complete treatment for potable water supply, the energy and resource requirements are typically less than would be required for conventional treatment technologies utilizing coagulation and filtration, chemical oxidation, carbon adsorption, or membrane filtration. This is because water treatment is primarily accomplished by alluvial materials that utilize naturally sustaining chemical and biological processes along with physical filtering to produce potable water.

The treatment and conveyance of water consumes a large amount of energy. In the United States, typically 20% or more of a communities' total energy use is associated with water treatment, conveyance, and delivery (Klein et al. 2005). Consequently, energy efficiency related to the production and treatment of water can result in reduced costs in addition to reduced environmental impacts such as greenhouse gas (GHG) emissions.

Compared to surface water abstraction, more energy is needed for abstraction of bank filtrate and groundwater due to the resulting drawdown in a well during pumping. Despite this, the energy requirement for natural filtration processes are typically less than other treatment systems since there is essentially no additional energy required for the treatment process. Other treatment technologies require significant energy to treat potable water due to the need to move water through a system of tanks and clarifiers, pump water through membranes or filters, and backwash filters. Natural filtration systems can accomplish the same task without this additional energy requirement.

The chemical requirements for natural filtration systems are also typically low since there is no need to use coagulants or other chemicals related to the treatment process. For natural filtration systems that provide complete water treatment, disinfection may be the only additional treatment process required. Similarly, there is little or no waste stream generated by natural filtration systems as there are no chemicals introduced into the treatment process.

In addition to the operational efficiencies, the resources required for the construction of RBF facilities are often less intensive in terms of the amount of materials and resources required for construction. For example, wells and associated equipment typically require less steel and concrete to construct than other types of potable water treatment systems.

5.2.2 *Other Environmental Advantages*

RBF facilities can be constructed and operated in a manner that has fewer impacts on the environment than traditional surface water treatment systems. For example, because natural filtration systems withdraw water from the subsurface, they are less intrusive to fish and sensitive habitat than treatment plants that rely on surface water intake structures. Also, the physical layout of RBF systems is usually much smaller than other treatment systems and can be more easily constructed to avoid or reduce impacts to sensitive terrestrial habitat.

RBF systems draw upon a combination of bank filtrate from surface water and land-side groundwater and thus can be managed to limit the abstraction of surface water during low flow periods of the river to ensure the ecologically necessary flow in the river.

From a sustainability perspective, another advantage of RBF systems is the lower amount of GHG emissions relative to other treatment technologies due to the reduced energy requirements for these systems. In addition, because of the reduced chemical usage and the lack of a waste generated, GHG emissions due to transportation of chemicals and waste products are significantly reduced. Finally, because there are no waste streams generated, the environmental impacts associated with waste disposal are avoided.

5.2.3 *Flexibility and Adaptability of System Operation*

The predicted effect of climate change for many areas of the world is a less consistent and reliable water supply due to increased seasonality and variability of rainfall events. Consequently, it may be important to divert and treat water when it is available, even if it is not fully needed at that time.

The use of natural filtration can be an important strategy to cost-effectively manage water supplies under these conditions because facilities are significantly easier and less expensive to construct and operate than other types of treatment facilities. For example, the costs to operate a well network are significantly less than a surface water treatment plant in terms of labor, chemical, and power costs. Also, it is much more cost-effective to respond to variable river flow conditions by operating such facilities during episodic high flow events than to respond to these variable conditions at a conventional treatment plant. Further, production capacity can be increased by operating infiltration ponds or seasonal dams to supplement infiltration to the aquifer during lower river flow events. Employing the adaptability and flexibility of natural filtration systems, communities can realize improved reliability and sustainability of water supplies during times of drought and changing weather conditions.

The inclusion of natural filtration systems into integrated water resource management is an example of adaptively managing changed or more variable water supply conditions to provide a more sustainable and reliable water supply. Figure 5.2 illustrates how a natural filtration system can be used as part of a conjunctive use program to utilize surface water when it is more plentiful (and the demand for water is low) so that it can be stored for subsequent use when surface supplies are lower (and water demand is high). This is also a common way to enhance economic expansion in water stressed areas. For example, the Flint River in southeast Georgia, USA is hydraulically connected to the bedrock aquifer. Irrigation pumpage reduces flow in the river during summer months. Georgia is not allowing new diversion of water for industries or municipalities from the Flint River. Currently, the state pays the farmers not to irrigate in drought years. One option to effectively manage the water supply needs for industries is to extract water from the Flint River during periods of high flows and storing that water underground in aquifers after necessary treatment. A major portion of the injected water can be pumped back to supply to industries (Sharma and Ray 2010).

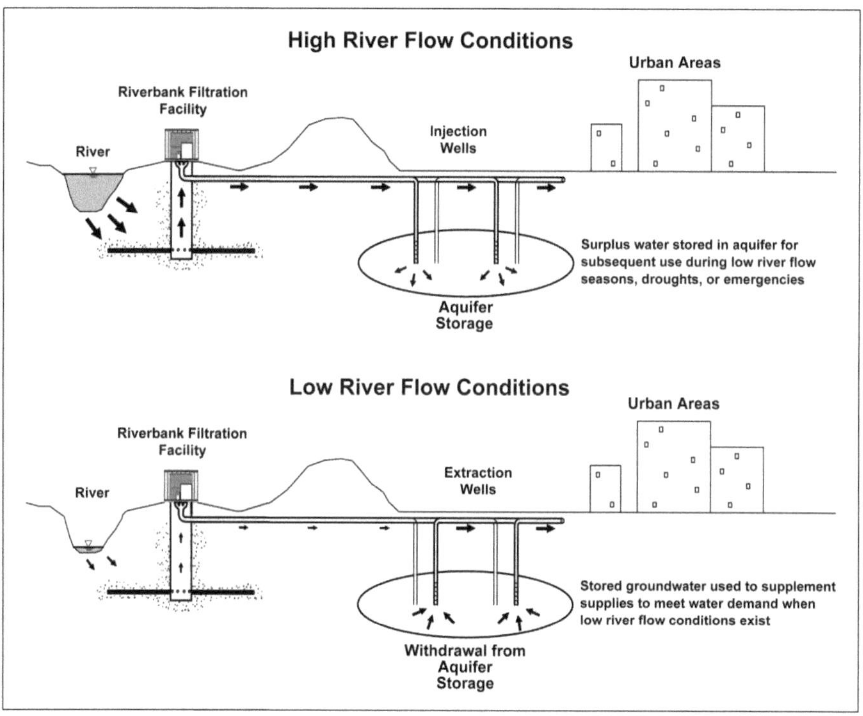

Fig. 5.2 Illustration of riverbank filtration facilities operated as part of a conjunctive use strategy to increase sustainability of water supplies during seasonal or long-term water shortages

5.3 How Does It Work?

The basic principle of bank filtration is induced infiltration from a surface water source. As stated earlier, if a surface water body is hydraulically connected to an aquifer and wells placed on the banks of the surface water body are pumped, a portion of the surface water is "induced" to flow to wells. As the induced surface water enters the riverbed and subsequently moves towards the well screen(s), suspended and dissolved contaminants are removed. Removal mechanisms for particles and pathogens (bacteria, viruses, protozoa) have some similarity with slow sand filtration.

Straining is a physical removal mechanism of particles when the diameter of the particle is larger than the grain size of the aquifer material (referred to as collector). Since the particles are unable to penetrate the granular media by virtue of their size, there is no Coulombic interaction between particle and grains. If the ratio of the particle diameter to mean grain size diameter is >0.005, straining becomes an important mechanism of particle removal (Bradford et al. 2003). However, ratio alone may not be an absolute parameter in determining straining. The angularity of grain size and heterogeneity in granular media could play a larger role in determining straining potential than that estimated solely by considering the ratios of particle to grains (Tufenkji et al. 2004).

When the available sites for attachment for the particles become saturated, the particles start to form bridges. This phenomenon is called 'ripening.' This typically occurs when the number of particles advected (moved with the flow water) through the porous media is higher than the number of sites available for attachment. As the filtration process proceeds, the new particle is unable to approach the collector due to geometric considerations and hence start to form a bridge (Elimelech 1998).

Physico-chemical filtration is the primary mechanism of removal of colloidal particles from surface water due to Coulombic interaction between particle and grain surfaces. The Coulombic interaction is the net sum of Derjaguin and Landau and Verwey and Overbeek (DLVO) forces acting between the particle and collector (here the aquifer sand). Derjaguin and Landau (1941) and Verwey and Overbeek (1948) together put forward an explanation to the mechanisms that determine whether a colloid will be transported or attached to the collector surfaces. This theory states that colloids will be mobilized or filtered depending on the sum of three types of interactions occurring simultaneously between the colloid and collector surface. They are the London–van der Waals (attractive) forces, the double layer (repulsive) forces, and "non" DLVO forces like hydration and steric interactions. Steric interactions are often controlled by the addition of polymeric substances to remove colloidal particles from surface water in coagulation/flocculation process in water treatment plants. Once the energy barrier between the particles is reduced and they collide, attachment and subsequent removal from the flowing water can occur.

Cake filtration is another mechanism typically observed in membrane filtrations in which the suspended particles form a clogging layer on the surface of membranes (Elimelech 1998). In the case of RBF, a clogging layer (analogous to the cake layer in cross-flow membrane filtration) can form if the suspended-particle concentration in the river is high and the infiltration rate is fast. For lakebank filtration, the deposited particles on the bottom of the lake may form a thick clogging layer.

Besides the above physico-chemical process, biochemical processes play a significant role in removing dissolved chemicals and pathogenic microorganisms. The surface charge of particles in the subsurface plays an important role in attracting and adsorbing colloidal particles and pathogens as well as in degrading certain chemicals. Under certain conditions, however, the surface charge can be altered or the magnitude reduced. Natural organic matter may coat the surface of metal oxides on grain surfaces, which reduces their ability to attract negatively charged particles or pathogens. Oxygen in the infiltrating water is consumed by microbes as they use the dissolved organic carbon (DOC) in the infiltrating water or use some of the particulate bound organic matter for metabolism. A reducing environment develops with the depletion of oxygen. As oxygen is consumed, nitrate present in the water acts as an electron acceptor in the redox reactions. With proper design and placement of wells, the surface water nitrate can thus be significantly reduced. If the wells are placed farther away from the banks (e.g., what is optimal for nitrate removal), Fe and Mn present in aquifer minerals subsequently act as electron acceptors. Fe- and Mn-(hydr)oxides will be reduced to Fe^{2+} and Mn^{2+} and are released in dissolved form. Sulfate reduction may occur in extremely reduced conditions.

Removal mechanisms for many chemicals present in surface water (e.g., pesticides and pharmaceutical compounds) are closely tied to the redox status of the subsurface. Water temperature and the amount of DOC are considered drivers of redox reactions. Certain pharmaceuticals (e.g., phenazone) are removed more effectively in winter conditions when the subsurface environment remains more aerobic due to the higher oxygen saturation of cold water and a temperature-influenced lower rate of oxygen consumption for DOC degradation (Massmann et al. 2008). The pesticide atrazine also degrades under aerobic conditions. There are other chemicals (e.g., explosives), however, for which removal is favored by anaerobic conditions.

Key design considerations in bank filtration systems are the sizing (length, diameter, open area) of the laterals and determining the number of laterals when a collector well is used. For vertical wells, screen design is the key. Spacing between wells and the offset distance from the water body are other factors to consider. Design and construction companies that have worked on wells for municipal and industrial water supply have developed empirical rules for entrance velocity through the well screens and axial velocities in the laterals. For vertical wells, an entrance velocity in the range of 1.83–3.66 m (6–12 ft) per minute is used over the screen zone. For collector wells, a design entrance velocity of 0.31 m (1 ft) per minute is used over the screen zone of the laterals. Because of slot blockage by sand and gravel, the actual flow velocity in the screen slots can be higher (as much as 2 times). The axial flow in the lateral pipe is limited to 1.52 m (5 ft) per second to reduce head loss.

5.4 Regulatory Perspective

5.4.1 United States

In the United States, the Long Term 2 Enhanced Surface Water Treatment Rule (LT2ESWTR, see USEPA 2006) dictates the level of treatment for a given source water. The LT2ESWTR was developed by the United States Environmental Protection Agency (USEPA) to improve drinking water quality from sources such as rivers, lakes, reservoirs, and aquifers by mandating removal of pathogens and other contaminants. Protozoa such as *Giardia* and *Cryptosporidum* are often found in source waters and can cause intestinal illnesses. *Cryptosporidium* is resistant to chlorine, one of the most common disinfectants for public water systems in the United States, and has caused outbreaks of parasitic disease cryptosporidiosis. LT2ESWTR targets additional *Cryptosporidium* treatment requirements to higher risk systems, requires provisions to reduce risk from uncovered finished water storage facilities, and provides provisions to ensure that systems maintain microbial protection as they reduce formation of disinfection byproducts (result of chlorination of drinking water). A reduced amount of chlorine or addition of combined chlorine (such as chloramine) may reduce the formation of trihalomethanes; however that may limit the inactivation of cryptosporidia.

USEPA defines groundwater under the direct influence (GWUDI) of surface water as the water that contains substantial amounts of "recent" surface water and is regulated as surface water. Microscopic particulate analysis (MPA, see USEPA 1992) is a consensus method followed by the USEPA to determine GWUDI classification. Contaminants and indicators examined include green algae, vegetative debris, rotifers, nematodes, crustacean and insect parts, diatoms, fungal spores and pollen, and inorganic particles. For GWUDI classification, the water often reflects the turbidity, temperature, specific conductance, and pH of the surface water (40CFR 141.2). The USEPA determines the additional treatment needs for water system (beyond current requirements) based on the concentrations of *Cryptosporidium* in source waters. It goes through four "Bin" classification systems in which Bin 4 requiring 2.5 additional logs of treatment (Table 5.1). The source waters need to be monitored at designated frequencies (which vary by the size of the water utilities) to estimate average numbers of cryptosporidia present in the water.

Table 5.1 Bin classification system and treatment needs for water systems

Bin number	*Cryptosporidium* concentration (in oocysts/l)	Additional treatment beyond current requirements
1	<0.075	No additional treatment
2	≥ 0.075 but <1.0	1.0 log
3	≥ 1.0 but <3.0	2.0 log
4	≥ 3.0	2.5 log

5.4.2 Europe

In Europe, there is no specific regulation for bank filtration sites. Based on long-term experiences and often larger distances between the river/lake and the wells in the United States, the common rules for the design of well-head protection zones are applied. The retention times of bank filtrate in the aquifer under European conditions (having mainly vertical wells) are mostly between 20 and 300 days, but can range in extreme cases from about five days to more than 10 years. In some countries (e.g., The Netherlands and Germany), a minimum travel time of 50 days is proposed because an old rule of thumb dictates that about 50 days are sufficient to obtain water free of bacteria. Problems with *Cryptosporidium* or *Giardia* are rare, so this is not the focus for regulation. If the travel time is more than 50 days (or the actual number of days typically used in a certain country), there is no problem for getting permission for a new RBF site if all other aspects related to quantity, ecology etc. are according to the rules. If the travel time is shorter (may be for only a small portion of the bank filtrate), the local authority will determine if the RBF facility to be sited at an alternative place to achieve the needed travel time or a permission is given for the RBF system to develop if appropriate water quality monitoring is assured.

Recent findings indicate that the riverbed infiltration surface area (and associated flowpath length), riverbed properties, and redox conditions play an important role in the removal of pathogens. Thus, travel time cannot be solely used as the single parameter for the evaluation of removal efficiencies and optimal operation of wells.

In Europe, concerns regarding bank filtration focus on the removal efficiency for organic micropollutants (e.g., pharmaceutically active compounds, endocrine disruptors) and the necessity for further treatment (activated carbon filters, nanofiltration). Such concerns have not yet resulted in any final regulatory measures. RBF along major rivers has, however, been an important driver for improved river sanitation and cross-border cooperation of waterworks and authorities (e.g., along the Danube, Elbe, and Rhine rivers).

5.5 Key Planning Considerations

Multiple factors must be considered in evaluating whether RBF is a viable water-supply technology for any new site. Clearly defined project objectives are essential prior to any significant commitment of resources. It is prudent to phase planning activities so that each phase builds on the prior, starting with lower-cost activities and progressing towards more complex and expensive activities. This section addresses important factors in evaluating the probable success in the use of RBF at a given site.

5.5.1 Define Project Objectives

Especially given the inherent uncertainties associated with implementing RBF at a new site, specific objectives for the proposed RBF system must be clearly defined early in the planning stages. Key considerations include:

- What is the end use of the water (e.g., drinking water, agricultural irrigation, industrial process water)?
- What level of treatment is required for the water (complete treatment or pre-treatment)?
- What is the demand for the water that must be met and how is that demand distributed throughout the year?
- Which portion of the river or catchment area along the river can be used to obtain water?
- What amount of water may be withdrawn from the river on a continuous basis without harmful ecological effects?

Given the inherent uncertainties regarding the effect of environmental factors on the operation and performance of a proposed RBF system it is important to continuously review all newly acquired site data and examine the significance of such new data relative to the project's objectives. Throughout the planning process policy/decision makers should be informed of the risks and uncertainties in meeting project goals so that the project may be modified as warranted by new information.

5.5.2 Phased Planning Process

The primary goal of an RBF assessment is to develop an understanding of the range of spatial and temporal variability that should be anticipated in system design. Given the complex relationship between surface water and groundwater it is prudent to conduct a phased investigation so that information obtained in early stages guides subsequent characterization activities.

Initial efforts should focus on reviewing all available existing information concerning fluvial processes, hydrology, and river geomorphology. Key information sources include any available flow records, historical mapping, historical and current photographs, land-use records, riverbank and riverbed survey data, and water-temperature data. After the review of existing information a site-assessment program should be designed that employs field-data collection and testing methods based on the objectives and resources of the project.

5.5.3 River Hydrology

River hydrology is an important consideration in selecting an RBF site. The hydrology of the river strongly influences surface water and groundwater interactions because the occurrence and distribution of high flow events influences the depositional environment along the riverbed and larger-scale fluvial processes of the river system.

For example, many rivers in the western United States exhibit a wide range of flow rates throughout the year, typified by high-magnitude flow events caused by winter storms or spring snow melt as well as short-duration relatively low-flow conditions in the summer months. River systems experiencing a wide range of hydraulic energy throughout the year will likely exhibit a high degree of depositional variability along the riverbed. They will likely also exhibit potential channel instability leading to increased spatial and temporal variability in surface water and groundwater interactions. Periodic high-magnitude flow events can cause riverbed scour and the removal of any clogging layer that may have developed from earlier depositional process. Conversely, rivers with more consistent hydrographs are usually characterized by less variability in riverbed depositional environment and fluvial processes, resulting in less variability in surface water and groundwater interaction.

The runoff regime of a river may initially be characterized by the average monthly discharge developed from daily data over several decades. Averaging discharge or level data is helpful to understand the basic properties of the runoff regime. But only unaltered original data give insight into the runoff dynamics of a river (Grischek et al. 2007).

Human alteration of stream channels (e.g., for flood control, hydroelectric generation, and/or improved navigation) changes the natural runoff regime. Dams dramatically alter the flow characteristics of rivers—particularly groundwater interactions, riverbed erosion, and the transport and deposition of solids.

5.5.4 Fluvial and Geomorphic Processes

Surface water and groundwater interaction is affected by changes to the river system brought about by fluvial and geomorphic processes. It is important to understand the river geomorphology in conjunction with its hydrology. Potential RBF sites should be evaluated with an understanding of fluvial processes and their ability to cause instability to the river system (Grischek et al. 2007). River instability can result in lateral migration of the channel route. For example, RBF facilities are often located along river bends, which can be prone to channel migration, dramatically changing the location and extent of river recharge to the underlying aquifer. Instability of the river system can also result in vertical changes to the riverbed slope through aggregation or degradation processes (possibly caused by local structures such as bridges or flow-diversion structures). These changes can affect surface water and groundwater interaction by changing the composition of the riverbed in addition to influencing water quality by decreasing the thickness of the aquifer.

The development of meanders in flat-country regions is a natural phenomenon of river channels caused by the inertia of flow. In meanders or bends, the maximum flow velocity occurs towards the outer section of the bend. This part of the riverbed is washed out by erosion and the riverbed is deepened, creating an asymmetric cross section. In this dynamic process, the larger fraction of the riverbed material that resists erosion (e.g., cobbles and pebbles) accumulates and may cause a paved bed along the outer section of the bend. The resulting riverbed is immobile even during the passage of flood waves (Schubert 2002). If the riverbed is paved or armored by cobbles and pebbles then induced infiltration of river water can cause filling of the spaces between the cobbles with very fine material (suspended solids) that cannot be removed during floods. In such cases, the permeability of the riverbed decreases dramatically. In regions near the inner section of the bend, the flow velocity is lower and alternating deposition and erosion may occur—the riverbed material remains movable. Meanders of the river channel are preferred sites for RBF plants if the proportion of bank-filtered water in the extracted well water should be high (Grischek et al. 2007).

Caldwell (2006) documented slopes for several streams at RBF sites in the United States and Europe, indicating that most surface slopes on unimpounded rivers ranged from 0.2 to 0.8 m/km. About 90% of the RBF facilities along the German part of the Rhine River are situated in the Lower Rhine region with a hydraulic gradient of the river level of 0.17–0.23 m/km, average shear stress on the riverbed of 12–8 Newton per square meter (N/m^2), average grain size of bed load of 8–13 mm, and average grain size of the riverbed of 14–32 mm. At many RBF sites in Europe, the average flow velocity of the river is more than 1 m/s and the average shear stress is higher than 5 N/m^2 (Grischek et al. 2007).

5.5.5 Watershed Conditions

It is important to understand the potential effects of land use within the watershed on the water quality of the surface-water resources. A source-water assessment should be conducted in the early stages of planning an RBF facility to inform the development of a water-quality monitoring program. The survey should include an examination of land uses that identifies potential point sources of water contamination such as agricultural, industrial, or wastewater discharges as well as non-point sources. In addition, historical land uses that are no longer operational but still may have an impact on water quality, such as past industrial or mining operations, should be identified. The assessment can also identify engineering measures (e.g., well-construction measures such as surface seals and/or well setbacks) or operational strategies to minimize exposure to potential contamination.

A history of intensive agriculture in Europe has caused an increase in nitrate and sulfate concentrations in the groundwater there, especially in the vicinity of large cities. Furthermore, waterworks built a century ago outside the existing urban area are now located within the enlarged urban area and face problems from urban groundwater pollution and a wide variety of contaminated sites. If the existing wells continue to be operated then additional treatment is often needed.

Another watershed factor that should be considered is whether or not there are sensitive species within the watershed and whether existing habitat protections for sensitive species may limit construction of facilities within the riparian zone.

5.5.6 *Surface and Groundwater Quality*

The quality of the available surface water will greatly affect the operations of an RBF system. Therefore a comprehensive water-quality monitoring program is an essential component of the planning process. A first step is to review land-use practices within the watershed to distinguish between anthropogenic and natural sources of water-quality impairment as described in Sect. 5.5.5. Based on this assessment a surface water and groundwater quality-monitoring program can be designed to account for specific watershed conditions. In general a surface-water-quality monitoring program should evaluate:

- Water-quality parameters that will affect facility operations including dissolved oxygen, DOC, hardness, pH, specific conductance or chloride, nutrients, minerals, temperature, total dissolved solids, total organic carbon (TOC), and turbidity
- Anthropogenic contamination such as healthcare products (including pharmaceutical chemicals), volatile and semi-volatile organic compounds, and wastewater compounds
- Microbial contamination by bacteria, cryotosporidia, giardia, and viruses that may be more episodic in occurrence than other constituents of concern
- Other potentially harmful contaminants from anthropogenic or natural sources (e.g., metals such as methyl mercury or lead)
- Algae and related malodorous and toxic compounds such as microcystins

The seasonal variability of these constituents should also be understood. Although RBF can provide an advantage in terms of buffering seasonal variations of surface-water quality, such surface-water quality variations can negatively affect water quality or reduce the reliability of RBF operations. For instance, some rivers or streams may exhibit a relatively high contribution of wastewater during low-flow seasons, a time when water demand (and therefore pumping) is at its highest level. In addition some contaminants, such as pathogens, may be episodic in nature.

An assessment of groundwater quality is important to identify if there are any constituents (e.g., arsenic, heavy metals, iron, manganese, nitrates, sulfates, and/or volatile organic compounds) occurring in the groundwater that will require treatment in addition to RBF. An assessment of groundwater quality will also be important to evaluate whether the system will be prone to operational difficulties such as biofouling.

5.5.7 *Water Temperature*

The temperature of river water can also influence the rate of infiltration to the subsurface (Constantz et al. 2002; USGS 2003). Hydraulic conductivity is a function of fluid and material properties. Because the viscosity of water is directly related to temperature, changes in water temperature alone will result in changes in infiltration rates along the riverbed. Assuming all other factors remain constant, warmer water will exhibit a higher infiltration rate than colder water. Temperature variations are observed on a daily and seasonal basis and can affect infiltration rates for both of these time scales. The affect of temperature on infiltration rates has been evaluated at the RBF facility at Louisville, Kentucky, USA (Hubbs et al. 2006) and Sonoma County, California, USA (Su et al. 2004).

A constant river-water temperature throughout the year in the same range as the preferred drinking-water temperature (e.g., 10°C) would be the ideal case. Whereas a constant temperature is a more theoretical case, rivers with a moderate seasonal change in water temperature may be preferred over those with strong temperature changes—which may cause gas formation and precipitation processes that affect the clogging of riverbeds. If there are strong water-temperature changes ($\Delta T > 20°C$) additional negative effects such as changing redox conditions or breakthrough of easily biodegradable compounds may occur. Such effects are reduced at RBF sites with a longer distance between the river and the wells or sites with a thick aquifer due to equilibration of water temperature in the aquifer and mixing processes (Grischek et al. 2007).

5.5.8 Geology and Hydrogeology

The geologic setting of a site strongly influences the suitability of the site for successful RBF. Alluvial and glacial deposits are generally more conducive to effective RBF systems than other geologic regimes. This is because these geologic deposits are comprised of loosely consolidated materials that create a tortuous pathway for water to travel from the river to the well screen, thus promoting biological degradation, chemical transformation, and physical straining of impurities. Other geologic regimes such as karst or bedrock formations are less effective than alluvial deposits in terms of providing high-quality water because water can "short circuit" the RBF process through fractures or large channels without the benefit of soil grain and water contact to promote effective degradation and of straining to remove impurities.

It is also important to recognize whether there are faults or other structural features that can create barriers to groundwater movement. These features may not be apparent unless water-quality testing or aquifer testing is conducted. Additionally, geothermal processes may affect water quality, possibly contributing undesirable or even hazardous concentrations of metals (e.g., arsenic) or minerals (Webster and Nordstrom 2003).

Localized geologic and hydrogeologic properties that can greatly influence the suitability of a site for RBF include:

- *Grain size, distribution, and hydraulic conductivity*—In terms of yield, coarse-grained alluvial materials exhibiting high hydraulic conductivity are most suitable for RBF systems. Most RBF systems operate at sites that are comprised of these materials having a hydraulic conductivity in the range of 10^{-2}–10^{-4} m/s. From a water-quality standpoint, poorly sorted granular materials may provide a higher degree of natural filtration than well-sorted materials because a more complex network of pore spaces is developed, enhancing physical straining and soil-grain-to-water contact.

- *Depth and extent of alluvial deposits*—Deeper and more extensive alluvial deposits result in greater groundwater storage and longer flow paths from the river system, which can mean higher-water quality from an RBF system. In general the thickness of the aquifer should be greater than 10 m (Grischek et al. 2007).
- *Interaction between alluvial aquifer and regional aquifer*—Most RBF systems operate in shallow alluvial deposits under unconfined aquifer conditions. The surrounding regional groundwater aquifer may represent a different groundwater regime (e.g., confined aquifer and/or reduced redox conditions) and affect the boundary conditions of an RBF site.
- *Grain surface chemistry*—As described in Sect. 5.11.1, grain surface chemistry appears to play a significant role in the effectiveness of natural filtration.

5.5.9 Composition of the Riverbed Hyporheic Zone

The composition of riverbed materials within the hyporheic zone controls the permeability and therefore strongly influences the hydraulic connection between the river and groundwater. The composition of material comprising the aquifer and riverbed are heterogeneous reflecting changing fluvial conditions during different alluvial depositional events (Grischek et al. 2007; Gorman 2004). For example, the composition of material along the riverbed coincident with the thalweg channel (exhibiting higher river velocities) is likely to be comprised of relatively coarse-grained material and more conducive to infiltration of surface water than finer-grained material deposited along the lower-velocity areas of the channel. Consequently it is important to understand how the local flow conditions (velocities) can affect the composition of the riverbed (Hubbs 2006).

In contrast to the underlying alluvial aquifer, the composition of the hyporheic zone continues to change temporally as a result of variable fluvial conditions. For example, during periods of high river flow the riverbed can be subjected to scouring conditions, removing the shallow, fine-grained sediments and leaving relatively coarse-grained materials. During periods of low-flow conditions deposition of fine-grained organic material occurs, reducing the hydraulic conductivity of the riverbed and affecting the rate of recharge of surface water to groundwater.

No general RBF siting advice can be given based solely on a sieve analysis of riverbed material (Grischek et al. 2007). The grain-size distribution of the riverbed material has to be evaluated together with river-hydrology data and shear-stress data. A useful tool is the Hjulström diagram which comprises the conditions for deposition, erosion, and transport (Hjulström 1935). Uniform grain size of riverbed material is inappropriate for RBF schemes, e.g., riverbeds armored with large pebbles are not scoured during floods and experience extreme clogging. If the grain size is mixed, the tendency for bed material to be moved during floods will increase, thus reducing clogging.

In addition to seasonal variations, clogging of the hyporheic zone and aquifer can occur over longer periods of time for operating RBF systems sited in loose unconsolidated alluvial material. Operation of RBF systems can create unsaturated and negative pressure conditions within materials beneath the river which can, over time, result in compaction of riverbed and aquifer materials thus decreasing the hydraulic conductivity (Hubbs et al. 2006; Su et al. 2007).

5.5.10 Surface Water—Groundwater Interactions

Interactions between surface water and groundwater are among the most important considerations when evaluating site characteristics for a possible RBF system. The relationship between the river and the underlying aquifer can affect both the yield and quality of water produced by RBF facilities. High-capacity RBF systems are typically located adjacent to rivers that exhibit efficient recharge of surface water to the aquifer. Even for situations where recharge from the river to groundwater is less than ideal, properly designed RBF systems can operate effectively and can realize a benefit of improved water quality due to lengthened travel time in the subsurface. Surface water/groundwater interaction at various geomorphologic settings, from the headwaters of a river to its confluence with lakes and/or oceans, is discussed in Grischek and Ray (2009).

For a river or stream environment, the interaction between surface water and groundwater is often described in terms of losing or gaining conditions (Fig. 5.3). A river that receives inflow from surrounding groundwater is referred to as a gaining river whereas a losing river exhibits a loss of surface water via infiltration through the riverbed and bank. Rivers can exhibit both gaining and losing conditions in different reaches or at different times of the year within the same reach.

An important feature controlling the interaction between surface water and groundwater is the active layer of riverbed sediments in which regular exchange

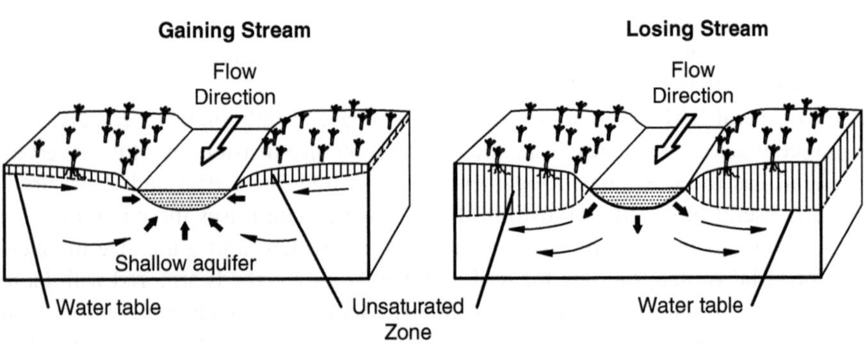

Fig. 5.3 Schematic showing gaining and losing stream conditions. (From USGS 1998)

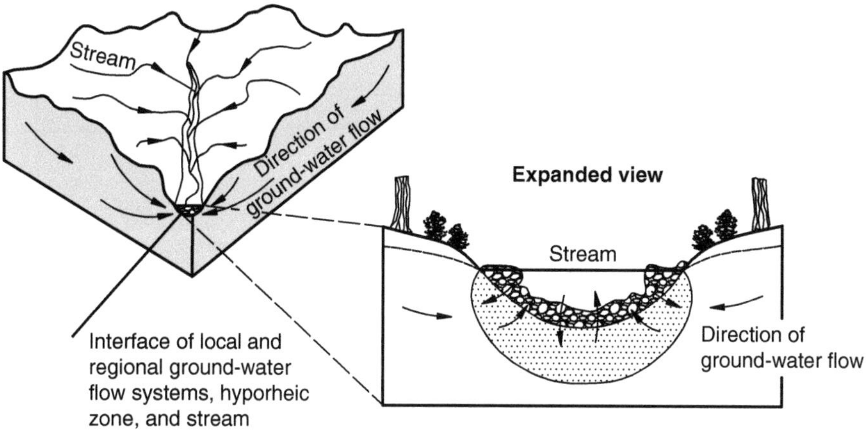

Fig. 5.4 Schematic showing the hyporheic zone within riverbed and bank sediments where active exchange of surface water and groundwater occurs. (From USGS 1998)

occurs between surface water and these sediments (Fig. 5.4). The hyporheic zone is often comprised of fine-grained, highly organic materials with relatively abundant nutrients, which provide a food source for a wide variety of microorganisms (Borchardt and Pusch 2009). As a result this zone often not only serves as the limiting or controlling feature governing the rate of exchange between the river and the underlying aquifer but also plays a significant role in improving water quality. Improved water quality is achieved because of relatively longer travel times, combined with finer-grained organic material and biological activity (Bencala 2000). A more precise definition of the hyporheic zone that translates to distinct identifiable boundaries in the field is difficult given the continuum of complex processes involved. More information can be found in Winter et al. (1998) and a special issue of *Advances in Limnology* edited by Borchardt and Pusch (2009).

Gaining at least a general understanding of the complex relationship between surface and groundwater is essential in the planning stages of an RBF facility. The spatial and temporal variability of the natural system, especially the hyporheic zone, makes this task complex and requires an understanding of the inter-relationship of several key factors.

5.5.11 *Natural Hazards*

From a system-reliability standpoint, it is important to assess the risk of natural hazards such as flooding or seismic events at the project site. It is almost always more cost-effective to incorporate design features in the original design that minimize risk rather than retrofitting existing facilities or, worse, repairing damaged facilities. Natural-hazard mitigation is a matter of risk tolerance, however, and project

Fig. 5.5 Floodwater covering the pump house of well 4 of Cedar Rapids, Iowa during the flood of 2008. (Photo from Roy Hesemann)

decision makers should be provided with information regarding the type of hazard, the relative risk, and the consequence of system impairment due to the identified hazards. The June 2008 flood of the Cedar River along Cedar Rapids, Iowa, USA, inundated the water wells (all 46 vertical wells and 3 of 4 horizontal collector wells) at the bank filtration plant located there. Two of the four horizontal collector wells had their motor platforms 0.75 m above the 500-year flood level. The flood level was more than 1 m above these platforms. Well 4 of Cedar Rapids was able to function because of sandbagging around the pump house platform and the availability of electric power (Fig. 5.5). This single well provided approximately 45,500 m^3 of water daily during the flooding to the citizens of the city and for emergency purposes.

In Torgau, Germany, an RBF waterworks at the Elbe River is abstracting up to 120,000 m^3/day. The mean discharge of the river is about 330 m^3/s. In August 2002 a so-called “flood of the century” occurred. The water level of the river increased within seven days by about 7 up to 9.47 m on 19 August, a level that had never before been observed. The peak discharge of the Elbe River in Torgau was estimated then to be 4,295 m^3/s.

The waterworks are located at an elevated level a sufficient distance from the river so that they were not affected by flooding. Nearly all production boreholes flooded but, because of their special watertight construction, they were still able to function. Concerns did arise with the power-supply system for the boreholes. While the transformer stations are located behind dikes, it was believed that the extreme high water level and the risk of dike failures mandated additional protection for the stations. Therefore the stations were surrounded by secondary walls of sheet piling and were provided with additional emergency-power generators and pumps for de-watering. As a result of these measures the power supply was maintained.

The flood caused a strong increase in organic carbon concentration, the number of microorganisms, and turbidity in the river water. The increase in DOC concentration from 5 to more than 10 mg/l in river water did not affect the DOC concentration in the pumped water. The increase in DOC in river water was mainly caused by an increase of the biodegradable fraction of the DOC. This was consumed along the flow path of the bank filtrate. Thus biodegradation and mixing within the aquifer prevented an increase in DOC concentration in the pumped water and any associated increase in disinfection by-products. Increases in concentrations of organic trace compounds such as pesticides, polycyclic aromatic hydrocarbons (PAHs), and organochlorines were not observed in Elbe River water during the flood. Despite many reported inputs of contaminants, the huge discharge caused an effective dilution and minimized the risk. Concentrations of contaminants after the flood were found to be around the mean annual values. Measured DOC and adsorbable organic halogens concentrations in bank filtrate samples were found to be within the long-term concentration ranges. Increased heavy metal concentrations in the river water also had no effect on raw-water quality. All measurements at the outflow of the waterworks proved that drinking-water quality was not at risk at any time during the flooding—the quality of the produced water continuously met the German standards for drinking water. Drinking-water disinfection was done with 0.2 mg/l ClO_2 and 0.6 mg/l Cl_2. There was no problem with bacterial contamination (Krueger and Nitzsche 2003).

5.6 Site Characterization

Considerable research is currently focused on the development and refinement of field methods to evaluate the relationship between surface water and groundwater. Given the inherent complexities there is now no single method or approach that is consistently superior to other methods for all sites. Consequently it is desirable to utilize a combination of methods with an understanding of the relative advantages and disadvantages of each method. The following summarizes some commonly used methods to assess surface water and groundwater interactions at potential RBF sites.

5.6.1 Riverbed Survey

A survey (bathometric or manual) of the riverbed can provide useful information when combined with flow records and soil-sampling results to gain an improved understanding of the locations of areas of relatively high infiltration. For example the thalweg channel may exhibit relatively higher hydraulic conductivities given the higher flow velocities and the composition of coarser-grained materials in those portions of the riverbed. In addition the slope of the river and bed structure (e.g., riffle/pool sequences) can provide useful information regarding river hydraulics and depositional environment (Hubbs 2006). Riverbed investigations using diving chambers were carried out at RBF sites on the Elbe and Rhine rivers in Germany (Schubert 2002; Heeger 1987). The results of a riverbed survey can assist in developing a sample grid for sediment sampling.

5.6.2 *Shallow Sediment Sampling*

Where river conditions allow, sediment sampling for analysis of grain size, organic matter, and microbial composition can assist in qualitatively understanding the composition and heterogeneity of the hyporheic zone (including riverbed sediments). For smaller rivers or streams, a sediment sampling program can be a cost-effective approach to gain data regarding the spatial variability of the hyporheic zone (Gorman 2004). For larger rivers, the logistics of collecting samples at depth under large river stage and high-flow conditions using diving chambers or freeze coring may prove cost prohibitive.

The results of a riverbed survey can assist in developing a sample grid. Sampling has to be done layerwise using core samplers. Hydraulic conductivity can be estimated based on grain size distribution using the Hazen correlation method (Fetter 2001), although systematic sampling errors caused by loss of fine grained material being washed away during sampling should be considered (Gorman 2004). The method of Beyer (1964) is also widely used. The method of Beyer includes not only the effective grain size diameter d_{10} but also the uniformity coefficient (d_{60}/d_{10}) and was developed for alluvial sediments. The hydraulic conductivity of the upper centimeter(s) of a riverbed can be up to two magnitudes lower than at greater depth. Determination of hydraulic conductivities based on sieving analyses is not reliable for samples from armored riverbeds or layers with a high portion of organic matter. Laboratory tests to estimate the permeability of sediment samples under constant and variable heads can also be employed but the previously mentioned sample bias should be considered. In addition, the results of grain size distribution analyses can be correlated to river flow to assess the potential for scour conditions (Hubbs 2006).

5.6.3 *Seepage Meters*

Seepage meters are commonly used to measure infiltration rates at various locations on the riverbed (Libelo et al. 1994; Rosenberry and LaBaugh 2008). These inexpensive devices can measure the rate at which water infiltrates into the river-

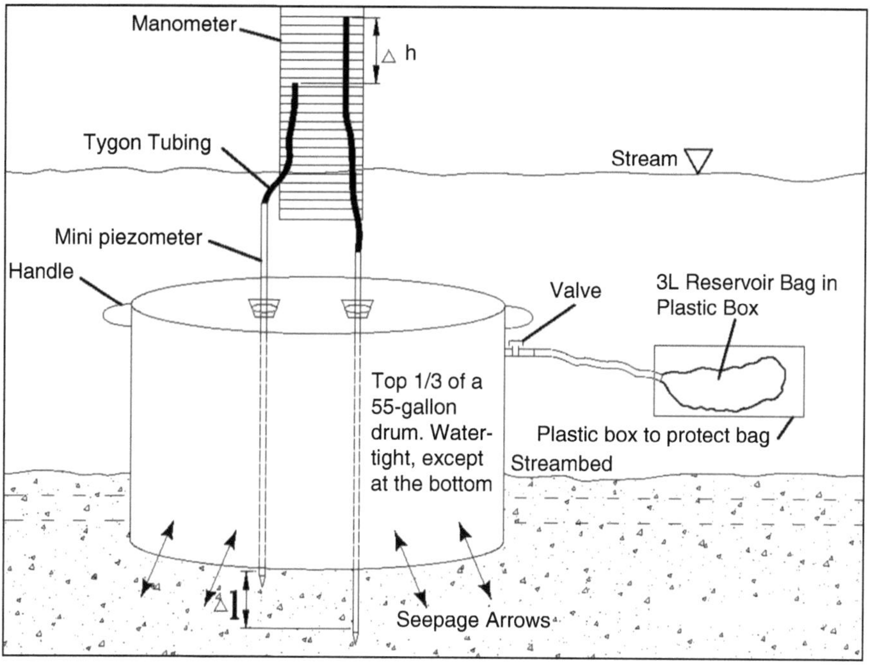

Fig. 5.6 Seepage meters are a relatively simple and low cost method of measuring infiltration rates along the riverbed. (From Gorman 2004)

bed. Figure 5.6 provides a schematic of a seepage meter. It is important to insure that leakage is minimized at the interface of the meter and the riverbed. This is especially important in coarse-grained materials. Seepage meters have been tested against seepage rates estimated using temperature measurements on the Russian River in California with good results (Cox et al. 2002). Seepage meters can be used efficiently in fine to medium grain size sediments but rarely in coarse sediments, armored riverbeds or beds including pebbles and boulders.

5.6.4 Analysis of Vertical Gradients

Measurement of water-levels in the river, and underlying groundwater provides information regarding whether the river is gaining or losing in the vicinity of the potential RBF site. The use of mini-piezometers (Rosenberry and LaBaugh 2008) driven into the riverbed at two to three nested depths (typically completed with 6-inch screened intervals) in conjunction with river stage can provide valuable data regarding the interaction of surface water and groundwater prior to site development (Rosenberry and LaBaugh 2008). Pressure transducers can also be used to monitor changes in river stage and subsequent responses at depth in the mini-piezometers to estimate vertical hydraulic conductivity of the hyporheic zone.

5.6.5 *Measurement of Water Temperature*

Temperature in surface water and in the shallow subsurface below the streambed can be measured to estimate the vertical hydraulic conductivity of the hyporheic zone. The potential for using heat as tracer was first discussed by Rorabaugh related to his work on an infiltration gallery along the Ohio River (Rorabaugh 1954). Heat is transported in flowing water, such that changing groundwater temperature patterns may be used to identify the rate and direction of groundwater movement to or away from a river (Rosenberry and LaBaugh 2008). Figure 5.7 shows conceptually how the hydraulic conductivity of the riverbed is determined using heat as a tracer for gaining conditions (Fig. 5.7a) and losing conditions (Fig. 5.7b). Temperature measurements can be obtained as part of the mini-piezometer program to assess vertical gradients. Figure 5.8 illustrates the correlation of water-level and temperature

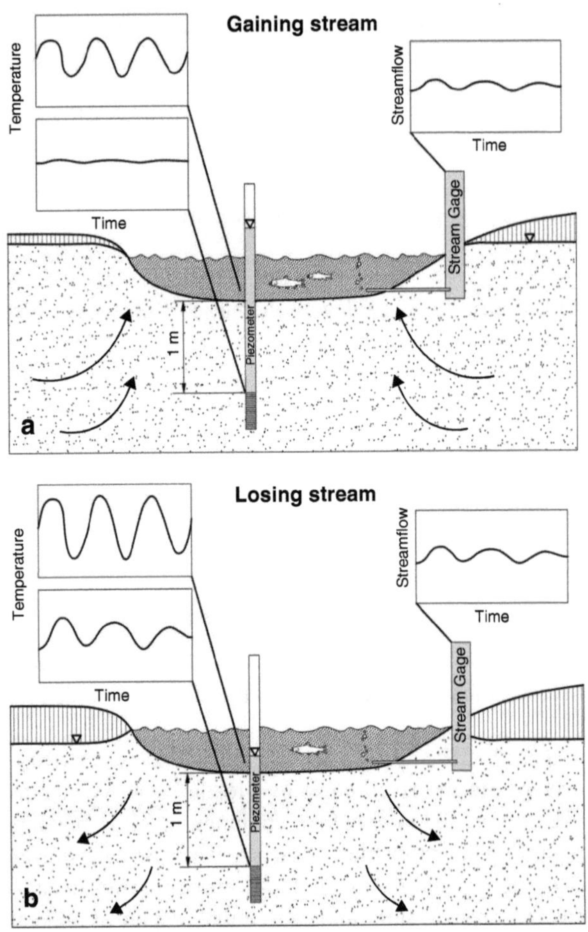

Fig. 5.7 Schematic illustrating how water temperature can be measured in surface water and groundwater to evaluate gaining and losing stream conditions: **a** gaining stream conditions do not exhibit transmission of diurnal temperature fluctuations from the river to underlying saturated media, **b** temperature fluctuations observed in the river are transmitted to the subsurface under losing stream conditions

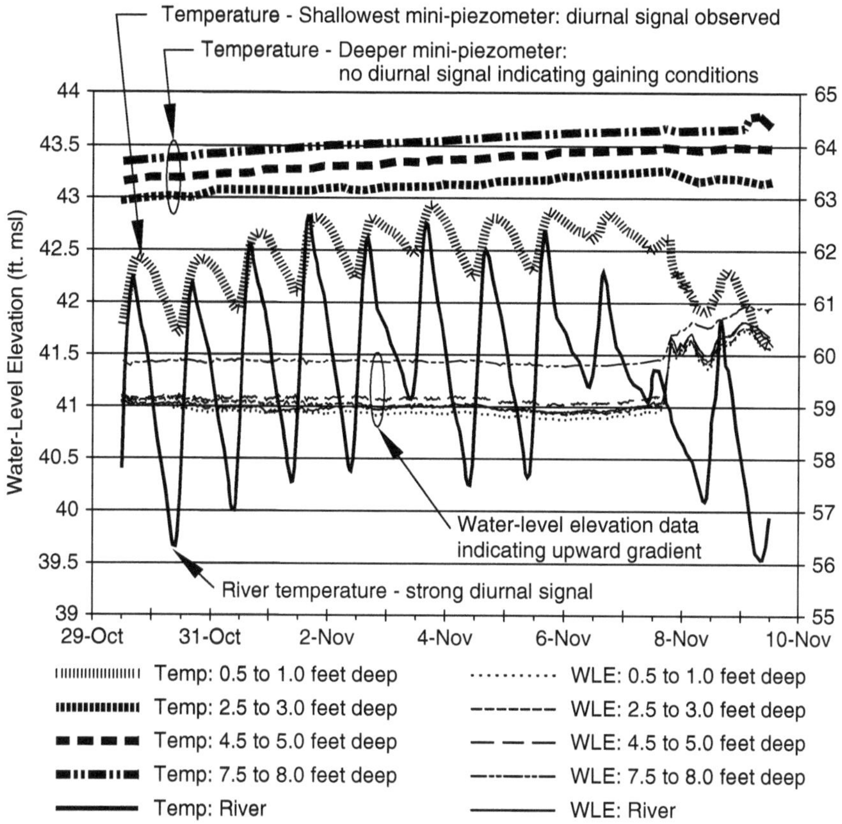

Fig. 5.8 Water level and temperature data from mini-piezometers from a site on the Russian River in California illustrating how these data can be used to evaluate surface water and groundwater interactions. In this case, both water level and temperature data indicate gaining stream conditions at this location

measurements at discrete depths to demonstrate gaining conditions along the Russian River in California (PES Environmental Inc. 2003).

The sediment temperature pattern over time, or sediment thermograph, may be used to inversely model heat transport and groundwater flow. For example, the model VS2DHI (Hsieh et al. 2000) was used to estimate the vertical hydraulic conductivity and water flux based on simulated temperatures in the riverbed sediments and calibration using observed temperature data. This is a form of inverse modeling that has proved successful in a range of groundwater modeling environments (USGS 2003; Constantz et al. 2002). Seasonal water temperature profiles in conjunction with water levels have been used at the Russian River in California to evaluate changes in riverbed conductance (Constantz et al. 2002; Su et al. 2004). The US Geological Survey Circular 1260 (Stonestrom and Constantz 2003) serves as a good reference regarding the use of heat as a tracer to evaluate surface water and groundwater interactions.

5.6.6 Seepage Runs

Seepage runs are often conducted to evaluate surface water and groundwater interaction on a larger scale (USGS 1969). A seepage run is essentially a surface-water balance. Seepage runs are conducted by measuring stream flow upstream and downstream of the potential RBF site and accounting for gains and losses in the river segment between the flow measurement stations over a fixed period of time. Longer seepage runs with more flow measurements can better account for system variations thus increasing the accuracy of the analysis. The boundaries of a seepage run need to be carefully selected to insure consistent flow measurements. Stream-flow measurements are made at locations that display the most uniform flow (e.g., straight reaches). Surveyed transects across the river are divided into several segments and velocity measurements are then taken along each transect at such locations. Gains or losses from springs, tributaries, or underflow are considered in addition to losses due to diversions, evapotranspiration, and uptake from riparian vegetation. After accounting for these factors the remaining gain or loss is attributed to surface water and groundwater interactions through the riverbed and bank. Increasing the number of flow measurements for a given section of the stream will improve the accuracy of the results and hence improve the understanding of surface water and groundwater interactions. This method is applicable only under optimal conditions and demands a high accuracy of discharge measurements, which is difficult to achieve for larger rivers.

5.6.7 Aquifer Testing

Hydraulic parameters of the aquifer where the wells are located need to be evaluated using pumping tests. These are standard tests where water from these wells is pumped at a steady rate and the drawdown of water level in the wells and in vicinity of the wells (e.g., at installed monitoring wells) is measured as a function of time. From the aquifer geometry, pumping rates, and time, hydraulic parameters (transmissivity and storage coefficient or specific yield) are determined. Although relatively expensive to conduct, an aquifer test also provides valuable information concerning surface water and groundwater interactions.

It is highly desirable to extend the duration of aquifer tests to allow the onset of river recharge because this boundary condition will strongly influence long-term operations. This can be accomplished by utilizing monitoring wells or piezometers located between the test well and the river, within the river (at different depths), and on the far side of the river (for smaller rivers). Monitoring locations beneath the riverbed, in particular, will allow for an assessment of whether unsaturated conditions develop beneath the river during the test.

Several analytical methods are available to estimate streambed conductance (Kruseman and de Ridder 1991; Kresic 1997). The results of these analyses can then be used as input for numerical groundwater-flow models to evaluate design scenarios. Aquifer testing is expensive because it involves the installation of a test well, several monitoring wells, is labor intensive, and requires the management of discharge water. Monitoring wells installed as part of an aquifer test program can later be utilized as a monitoring network for long-term operations.

5.6.8 Geophysical Methods

The application of geophysical methods is gaining interest for use in evaluating volumetric water content in the subsurface and infiltration rates (Hubbard and Rubin 2005). Surface electrical and ground-penetrating radar (GPR) methods have been used to quantify subsurface-water content and infiltration of surface water through porous media. Surface-based GPR involves emitting high frequency (50–1,000 MHz) electromagnetic signals into the subsurface. The signal is modified as a function of the dielectric constant of the soil. As the dielectric constant in soils is primarily governed by water content, analysis of the travel time of the recorded ground wave (Grote et al. 2003) and reflected wave (Lunt et al. 2005) can be performed to estimate soil water content. Because GPR is non-invasive and does not require installation of boreholes or wells it can potentially be a cost-effective approach to evaluating surface-water recharge. Significant limitations of GPR include its inability to be used directly over surface-water features (such as rivers), which limits use to riverbanks or to the edges of infiltration ponds, and signal attenuation in clay-rich environments, which limits the distance or depth over which the signal can propagate. An excellent review of GPR methods is given by Annan (2005).

Electrical resistivity is a measure of the resistance of materials to the flow of electricity; it is the inverse of electrical conductivity and is an intrinsic property of a material. Electrical resistivity measurement methods involve the introduction of a time-varying direct current (DC) or very-low-frequency (< 1 Hz) current into the ground between two electrodes. The currents can be envisioned to set up equipotential surfaces, with current flow lines running perpendicular to these surfaces. The fraction of current that penetrates to a particular depth is a function of the electrode spacing and the electrical resistivity distribution of subsurface materials. Modern electrical equipment now includes multiplexing capabilities and automatic and autonomous computer acquisition which greatly facilitate acquisition of pseudo-electrical-resistivity cross sections. While it has been successfully performed using site-specific controls (e.g., Sheets and Hendrickx 1995; Michot et al. 2003), because electrical conductivity is sensitive to many factors (e.g., moisture, temperature, soil texture, and pore fluid total dissolved solids) unique interpretation of electrical resistivity data in terms of moisture content can be challenging. Typically, surface electrical resistivity measurements are used to provide spatial patterns of electrical conductivity rather than to provide unique estimates of soil properties. An excellent review of the use of electrical resistivity methods for hydrogeological applications is given by Binley and Kemna (2005).

Tomographic radar is a subsurface method where electromagnetic energy is transmitted in one boring/well and received in another boring/well (Hubbard et al. 1997). Electrical resistivity measurements are also conducted between boreholes by inserting electrodes into a wellbore or installing electrodes on the outside of the wellbore casing and collecting measurements. This process is now commonly referred to as electrical resistance tomography (ERT). With both tomographic radar and ERT, inversion algorithms are necessary to reduce the recorded measurements into estimates of inter-well dielectric constant or electrical resistivity, respectively, which can in turn be used to estimate hydrological properties. Although wells must be installed, tomographic surveys can be conducted over shallow surface-water bodies. Such geometries may be best suited for detailed analysis of relatively small areas because the areal extent of the tomography is limited.

In addition to collecting surface and tomographic data at one point in time to characterize systems in a static sense, geophysical data can be collected at the same location over time. Observing the data as "time difference" sections (measurements collected at an earlier time subtracted from measurements collected at a later time), or "time-lapse" sections, enhances the imaging of subtle geophysical attribute changes. These data sets have been used to elucidate dynamic transformations, such as moisture infiltration, and thus have significant potential to assist in understanding hydrological processes. Surface electrical and time-lapse tomographic radar and ERT are all being evaluated by the Sonoma County Water Agency at an infiltration pond for an RBF site on the Russian River in California.

The measurement of streaming potential is another non-invasive geophysical method that shows promise in evaluating infiltration through porous media. When water, a polar compound, migrates through porous media with surface charge, an electrical current is generated. This phenomenon is referred to as streaming or self-potential (SP). SP has been utilized to measure seepage across earthen embankments (Sheffer and Howie 2003) and may be able to be used to qualitatively or quantitatively measure infiltration across a riverbed. A review of SP methods for hydrological problems is given by Revil et al. (2006). SP is currently being evaluated at a RBF site in Sonoma County, California (USA) to determine the viability of this method to characterize surface water and groundwater interactions (Gasperikova et al. 2008).

5.7 Design Considerations

5.7.1 Centralized or Decentralized Pumping?

An important design consideration is whether water extraction will be from centralized facilities such as radial collector wells where pumping is concentrated over a localized area or from a decentralized network of vertical wells that spreads the pumping stresses out over a larger area. Several factors must be considered in determining the best strategy for a particular site. Some of the relative advantages and disadvantages of centralized and decentralized pumping systems for a similar capacity system are discussed below:

- *The area available and extent of the aquifer*—vertical well fields typically require a larger areal footprint than radial collector wells for a similar capacity system.

- *The thickness of the aquifer*—if the thickness is minimal, horizontal laterals of collector wells are advantageous as vertical wells would have too short filter lengths.
- *The potential for aquifer compaction*—the higher rate more concentrated pumping typically used with horizontal collector wells can lead to a decoupling of surface and groundwater, causing compaction of the aquifer, thus reducing hydraulic conductivity and permeability of alluvial material near the collector well.
- *Pumping, pipelines, electrical, and appurtenances*—since collector wells concentrate the pumping facilities, there are usually fewer pumps, pipelines, electrical lines, and other appurtenances (such as meters and control equipment) than vertical well fields of the same overall capacity (except for siphon systems, to be discussed later).
- *Water quality*—given sufficient aquifer depth, horizontal intake laterals can make it possible to locate RBF facilities closer to a river and maintain adequate distance between the laterals and the river bottom to meet water quality goals, whereas vertical wells need to be located farther away from the river to achieve adequate separation between the river and the well screen. By using a collector well with laterals oriented parallel to the river, the infiltration and flow velocity along the river stretch towards the lateral can be equilibrated compared to flow towards vertical wells, thus collecting a lower proportion of water with short residence times.
- *Maintenance*—collector well maintenance can be less costly than a vertical well field of similar capacity because the operator needs to maintain less equipment. However, intake laterals of collector wells are always saturated, which can promote scaling due to changes in chemical equilibrium and fouling due to iron bacteria growth (Driscoll 1986). Remediation of laterals of collector wells is more difficult than for vertical wells.
- *Reliability and operational flexibility*—Because vertical well fields consist of multiple wells that are decentralized relative to a radial collector well system, they can provide more reliability in the event of equipment failure or a catastrophic event such as a earthquake or flood. Similarly, the operator has more flexibility in varying pumping rates and the location of extraction with a vertical well field than with a system comprised of radial collector wells.

5.7.2 Siphon Systems

Siphon systems are a variant of distributed abstraction systems having centralized pumping. A series of vertical wells are connected through a siphon pipe to a larger collector shaft. The siphon pipe is filled with water through the use of a vacuum pump. Thus the water levels in the collector and the wells are coupled and counterbalanced. If water is pumped from the collector shaft, the resulting pressure change in the pipe causes inflow of water into the vertical wells (Fig. 5.9). When designing the hydraulic system, well-yield calculations have to be coupled with pipe-flow calculations including flow-rate-dependent pressure losses and resistance coefficients. The system has to be designed carefully to prevent excessive drawdown in the wells

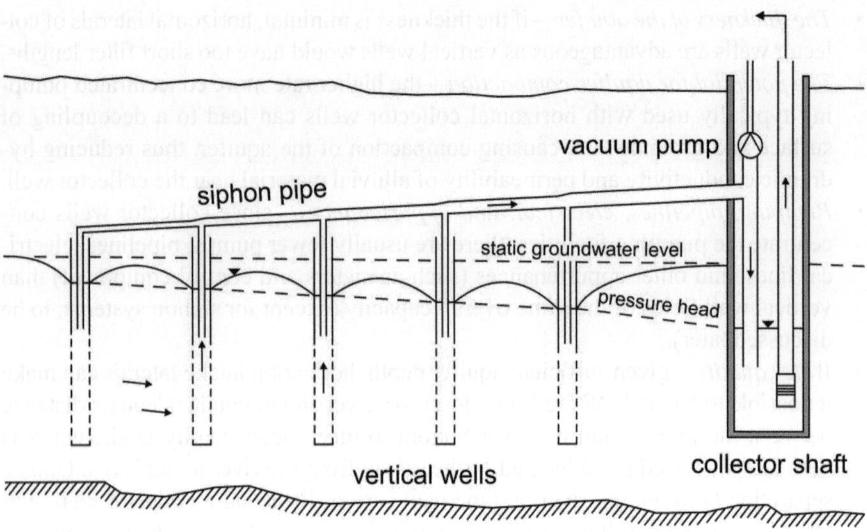

Fig. 5.9 A siphon system and its connection to a central collection caisson

that would allow air to enter the pipelines. The maximum practical drawdown is less than 7 m (the maximum theoretical drawdown is 10 m). Droughts and resulting low water levels have to be considered—as was done by Eckert et al. (2006) in discussing the impact of possible climate change on water levels and the operation of a siphon system at Düsseldorf, Germany. Construction specifications for siphon systems require the construction to allow no air entrance. Vacuum pumps normally still have to be operated on demand several times a day when air accumulates in pipes. Specific devices have been developed to minimize the need for vacuum pumping.

In Europe, siphon systems were built mainly in the beginning of the twentieth century at sites with an aquifer thickness of less than 20 m, when construction techniques of horizontal collector wells were not fully developed yet. Many systems have been in operation for more than 100 years (e.g. in Düsseldorf, Dresden, Karany, Budapest). Typically, the well yield and the infiltration rate per meter river length are low—such systems demonstrate RBF being used in a manner without excessive pumping. Advantages of siphon systems are low energy consumption and low maintenance needs, similar to horizontal collector wells. The systems are ideal for long term operation for basic supply. Compared to vertical wells with submersible pumps, more wells have to be drilled and the ability to react to higher demand is limited. Commonly, every single well can be decoupled from the system through a valve allowing single well maintenance and remediation. In older systems, single well yields are not known because flow meters were not installed to avoid pressure losses. Differences between wells are observed only by comparing water levels in the well and in the filter pack.

Figure 5.10 shows an example of a siphon system at the RBF water works Görlitz, Germany, along the River Lausitzer Neisse. Three well galleries with total 44 wells are connected to the old collector well, two well galleries with 32 wells

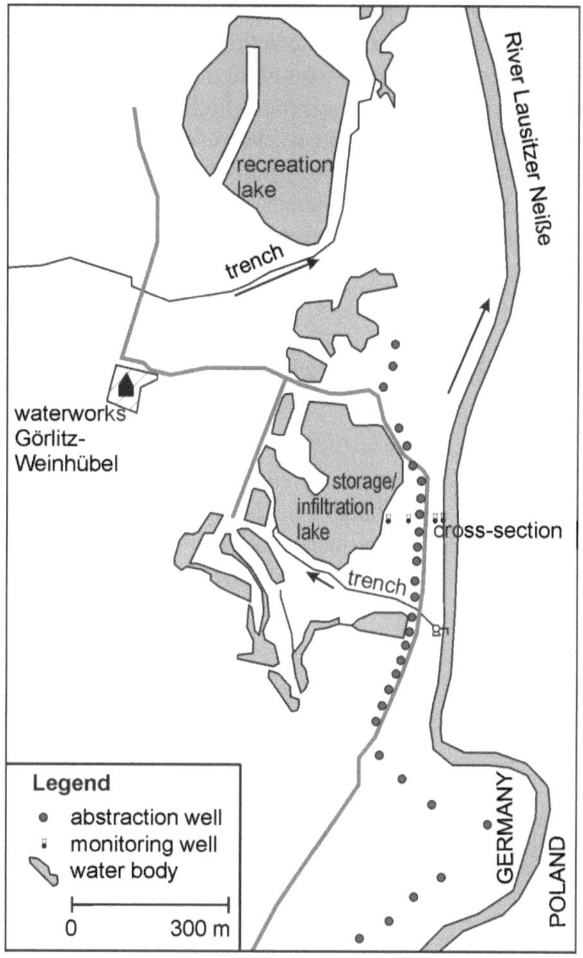

Fig. 5.10 Siphon system in Görlitz, Germany. (From Grishek et al. 2010)

to the new collector well. Each well gallery can be uncoupled from the collector wells. The depth of the wells is between 12 and 15 m. The capacity of the water works in 1970 was 22,000 m^3/day, resulting in single well abstraction rates of about 290 m^3/day.

5.7.3 Pump Selection

Selections of the type of pumps and the motor systems are important considerations in the design of RBF systems. Fixed-speed pump motors can be advantageous from a cost and maintenance perspective while variable-speed motors, offering significantly increased capability of flow control, are desirable from an operational perspective. The use of variable-speed pump motors to control flow to maintain relatively constant water levels in wells can help reduce scaling issues within the well screen by maintaining fully saturated conditions. Variable-speed pump motors can also reduce the potential for aquifer compaction (resulting in decreased aquifer hydraulic conductivity) caused by concentrated high pumping rates. Variable-speed pump motors, especially for larger systems, however, increase initial cost, energy consumption, and heat generation (possibly requiring an expensive cooling system for the pumps).

For smaller facilities variable-speed pump motors typically are desired. For larger wells and collector wells, fixed-speed pump motors are often used given concerns regarding cost, energy consumption, and temperature issues.

5.7.4 Enhanced Recharge Techniques

Various techniques are used in conjunction with RBF systems to enhance their production. Two common techniques used to enhance recharge of surface water to groundwater are inflatable dams and infiltration ponds.

An inflatable dam can be used to seasonally influence the stage of a river or stream (Fig. 5.11). The inflatable dam can be raised during periods of low river flow

Fig. 5.11 Inflatable dam on the Russian River in California, USA is used to increase production capacity of an RBF system by increasing groundwater levels upstream of the dam and allowing surface water to be diverted from behind the dam to infiltration ponds located adjacent to collector wells. Note water diversion and pumping facilities behind the dam and fish ladders on edges of the dam to provide for fish passage

to enhance RBF production by increasing river stage and wetted riverbed, thereby increasing groundwater levels. In addition, inflatable dams also increase the water level behind the dam so that surface water can be diverted to off-river infiltration ponds, which can be used to enhance RBF production.

Inflatable dams are more suitable to smaller rivers or streams that exhibit significant variations of stage levels due to variability of seasonal flows. Inflatable dams are installed in the stream channel with a bulkhead structure. The key feature of inflatable dams is the bladder system. A rubber bladder is installed within the bulkhead and is inflated with water or sometimes with air to raise the dam. Depending on the design, there is some ability to control the stage of the river by filling the bladder to various levels. For situations of low river flows in climates that experience significant diurnal temperature fluctuations, care must be taken to control or minimize the expansion and contraction of the bladder due to heating and cooling. Such temperature fluctuations can cause undesirable changes in the river stage and consequently flow downstream of the dam. The expansion and contraction of the inflatable dam due to temperature fluctuations can cause irregular flow downstream of the dam. In areas where sensitive fish species are present, the inflatable dam design will need to include fish ladders or other measures to reduce impacts to fish passage.

Infiltration ponds, often operated in conjunction with inflatable dams or other in-stream diversion facilities, can be used to effectively increase production capacity of RBF systems when natural stream flows are reduced (Fig. 5.12). It is important to have a means of removing sediment from the river water, such as a cleanable sedimentation basin, prior to conveying water to the infiltration pond via gravity flow in order to avoid excessive clogging of the infiltration pond. Where multiple

Fig. 5.12 Infiltration ponds along the Russian River, California, USA. Note location of collector well pump houses between river and infiltration ponds

infiltration ponds are constructed, the ponds can be operated on a rotational basis allowing periodic cleaning.

A key issue that controls the effectiveness of infiltration ponds is proper maintenance to reduce impacts from clogging. Prior to the season of use, the bottom of the recharge ponds should be cleared of silt or debris that may have been deposited during winter storm or flood events. In addition, vegetation should be removed from the ponds at all times. The optimal water level to operate a pond is site-specific and should be evaluated through system operations. Deeper pond water levels may reduce plant growth and the formation of organic material on the pond bottom, although there may be a greater driving head for the migration of fine grained materials deeper into the subsurface. Shallower pond water level may result in increased infiltration due to warmer water (reduced viscosity); however, clogging at the pond bottom due to plant growth or deposition of organic matter may reduce pond infiltration performance.

Pre-treated water (such as that coming from an infiltration gallery or a sedimentation basin) normally has lower turbidity. If this pre-treated water is used to infiltrate in deep basins, light can penetrate deep into the ponds inducing algae growth. The algae will grow and die with time and the detritus can eventually clog the bottom of the ponds. Consistently attentive maintenance is needed to address these various concerns.

Infiltration ponds and basins are used in Germany along the Elbe, Neisse, and Ruhr rivers. In the floodplain of the Lausitzer Neisse River, the Görlitz-Weinhuebel Water Works is located at the German-Polish border. A lake and several artificial infiltration basins are fed by river water to augment the quantity of available water in the abstraction wells (Fig. 5.8). This technique was established during the 1980s when water demand and consumption were at a relatively high level. However investigations in 2009 into the infiltration capacity of the lake found that the bottom of the lake was then almost completely clogged.

5.7.5 Lakes Application

Bank filtration systems on the banks of lakes have been operating for a long time in various locations in Finland (Miettinen et al. 1994, 1996); Berlin, Germany (Heberer et al. 2002a, b; Wiese and Nutzmann 2009; Gunkel and Hoffmann 2009); and Nainital, India (Dash et al. 2008). Unlike the flowing river, lake water is static. As a result, the deposited particles and organics often cause anoxic conditions beneath the lake bed. Lacustrine deposits reduce vertical hydraulic conductivity enhancing the development of anoxic conditions for the infiltrating water from the lake bed. Such conditions can lead to the dissolution of sulfate and iron. Proper siting of the wells can mitigate this. If this is not feasible, the dissolved iron (e.g., Fe^{2+}) and sulfide must be removed prior to distributing the water. Also, the DOC content of lake water can be higher than that of river water. The excess DOC can contribute to the development of anoxic conditions.

In developing countries such as India, cities are contemplating building RBF systems adjacent to river barrages (low dams). In the upstream sides of the barrages, the water is still (like a lake) and the riverbed can have sediments containing fine particles and organics. Such conditions can reduce infiltration as well as enhance the creation of anoxic conditions.

If a lake is located in glaciated areas with coarse bed material, clogging may not be a serious problem. If the water temperature is cold, the DOC degradation will also be slow, thus reducing the development of anoxic conditions.

5.8 Operational Considerations

5.8.1 Operational Criteria

The two major operational criteria that water utilities need to address are maintaining design yields during periods of peak demand and regulatory compliance of the filtrate quality. It is important to develop operations plans that are aligned with the water-production quantity and quality goals established during the pre-design stage when beginning the operation of newly constructed RBF wells. The water utility should have significant information regarding the prevailing quality of surface water and groundwater developed from monitoring during the planning process and other studies as discussed in Sect. 5.5.6.

The utility, in conjunction with appropriate regulatory agencies, determine if additional treatment is needed for the filtrate based on the performance of other operating wells in the area meeting the regulatory standards. Following such a review, if no additional treatment is needed besides disinfection, desired monitoring strategies for efficient operation of the system are prescribed.

5.8.1.1 Yield (Production)

The filtrate yield from a successful RBF system must accommodate daily and seasonal demand fluctuations on the water utility as well as seasonal and climate-related (e.g., drought) variations in surface water and groundwater availability. For example, California requires water systems to be able to meet 2.25 times the mean daily demand (see California Drinking Water Regulations, Sect. 64554[b][3][C]). Most water systems are designed to handle similar demand fluctuations. Note that daily demand spikes can usually be accommodated by system storage such as overhead tanks. Similarly the average day peak-month demand in summer can be twice as high as the winter demand. Although storage tanks can help mitigate short-term demand fluctuations the RBF system should be capable of meeting the yield requirements in periods of prolonged drought and seasonal increases such as summer irrigation needs.

Yield reductions are generally tied to (a) river stage, (b) aquifer/streambed clogging, and (c) well/lateral fouling. Aquifer clogging may result from chemical precipitation inside granular media or from biofilm growth but published literature does not address these issues. Aquifer clogging can result from chemical precipitation inside granular media when chemical conditions change, thus disrupting equilibrium (Driscoll 1986). This can occur in unconfined aquifers that undergo significant pumping and drawdown where atmospheric conditions can change the solubility of minerals such as calcium and magnesium resulting in mineral precipitation. Precipitated materials can result in a reduction in the porosity, hence permeability. Also, a reduction in porosity can be expected for conditions when fine particles enter the aquifer. Riverbed clogging is the main issue for RBF and is dependent on the dynamic action of the river, sediment size distribution and load, and the infiltration velocity at the river-sediment interface.

Well screen clogging is a common problem in areas having relatively hard groundwater. Chemicals such Ca^{2+} and Mg^{2+} are formed when conditions that can be triggered through chemical or physical mechanisms are favorable. While the chemical precipitation causing well clogging has been studied over a long time, physical effects, such as excessive pressure drop along the well screen or in the gravel pack (resulting in the release of carbon dioxide which can contribute to changes in water chemistry in the zone immediately around the well) have only more recently attracted attention. Biofouling of the well screen by iron bacteria can significantly reduce the production of RBF facilities. Iron bacteria growth is promoted when well screens are open to the atmosphere in environments with groundwater containing iron, manganese, and higher amounts of DOC (Driscoll 1986).

In Germany, before 1990 the planning and construction of water works and well galleries was based on a predicted water demand of up to 200 l per capita per day. Since 1990, many waterworks, especially in the area belonging to the former East Germany, have been facing drastic reductions in water consumption. Mean water abstraction rates for public water supply decreased due to the "water price shock" after the re-unification of Germany, demographic changes, and changes in consumption patterns. In some regions, water use decreased by more than 50% within 10 years. Given an expected decline in German population, a further decline in water consumption and thus water production is expected. At many German bank filtration sites, reduced water abstraction results in a lower portion of bank filtrate in the abstracted raw water. At all sites with long flow paths, mixing ratios of bank filtrate and groundwater were found to be of primary importance to the concentration of DOC, nitrate, sulphate, dissolved iron and manganese in the abstracted raw water. Thus, the quality of the landside groundwater becomes more important for the subsequent water treatment (Grischek et al. 2010).

5.8.1.2 Water Quality

Major considerations related to water quality at RBF sites include changes in river-water quality, changes in watershed conditions, and regulatory/permit compliance.

Operational monitoring of water-quality parameters is performed to meet regulatory requirements but additional monitoring is also often conducted to meet requirements of the utility. Regulated water-quality parameters in the finished water are monitored by the water utilities. The frequency and number of samples depends on the size of the system and the requirements imposed by the regulating authority. Additional monitoring of the water quality of the river (including the side of the river opposite the collector wells) and groundwater for possible adverse changes in water quality that could affect RBF operations is strongly recommended.

In the United States, a significant consideration that controls the regulatory requirements for individual RBF systems is whether the facility is determined to be using GWUDI. The Enhanced Surface Water Treatment Rule, administered by the USEPA dictates criteria for pathogen removal (Anonymous 2006a). Under this rule, USEPA provides treatment credits to entities that use RBF for treatment of water. For example, one log treatment credit means if a utility is required to remove 99.99% of a pathogen at a treatment plant, it can remove 99.9% if RBF is used as a pre-treatment before the water is treated at the plant. This regulatory determination will not only dictate the amount and type of sampling and analysis required for operating the system but will also control whether post RBF treatment is necessary (see Sect. 5.4)

5.8.2 *Operational Parameters*

As part of the operational program of an RBF system, several parameters must be consistently evaluated to meet the water-production and water-quality operational criteria described above.

5.8.2.1 Pumping Rate

Achievable pumping rate is one of the key operational parameters for the sustainable operation of RBF systems. Maximum achievable pumping rate is controlled primarily by changes in the natural environment in which the RBF system operates. For example, infiltration along the riverbed may vary seasonally depending on the natural setting of the RBF site. Some systems will experience lower surface-water infiltration caused by the higher viscosity of colder water during the winter (Constantz et al. 2002; USGS 2003) whereas other sites will experience lower surface-water infiltration caused by the deposition of fine-grained organic material on the riverbed during the summer (Su et al. 2004). Hydrology is also significant in determining when high and low flows occur and riverbed-scour events take place. The achievable pumping rate must be able to consistently meet the necessary production requirements regardless of seasonal water-demand fluctuations.

RBF provides an opportunity to employ conjunctive-use strategies (e.g., aquifer storage and recovery and groundwater banking) for situations where low river flows make it difficult to meet increased summer demand. Given the relatively low cost of operating RBF facilities, it can be advantageous to operate at a higher capacity than needed during seasons of high water availability (and low demand) for subsequent conveyance to and storage in other groundwater basins. The stored water can be subsequently used to meet short-term peak demand when river flows are low or on a longer-term basis for drought protection.

5.8.2.2 Frequency and Duration of Pumping to Meet Demand

Pumping rates from RBF wells will generally vary depending on demand. Most large collector wells have multiple pumps with one or more pumps operating at any given time. Large systems generally operate throughout the year. As water demand is generally low in winter in some instances wells are pumped at lower rates or pumping at some wells in a well field may be shut down. Often some wells will be recognized by utility personnel as "good producers" and are pumped more frequently than others as long as the water quality in these more-frequently pumped wells continues to meet regulatory criteria.

It is desirable to operate wells under the most consistent pumping conditions possible to maintain a steady water level in the well. For smaller, lower capacity wells, variable-speed motors can be effective in adjusting pumping rates to maintain consistent water levels in the well. For collector wells, variable-speed motors may be prohibitively expensive to operate. Adjustments for these systems may be made through the use of outflow valves or by using multiple pumps with differing motor capacities to provide operational flexibility.

5.8.2.3 Control of Stream Flow Releases/River Flow

The operation of an RBF facility must be coordinated with river-flow conditions (natural and regulated) to ensure production-capacity and water-quality goals are met. Water levels in the rivers can affect the quality of filtrate as well as the production quantity. In unregulated rivers, filtrate generated by a given drawdown during low-flow conditions can be of lower quality than that during high-flow conditions. Low-flow conditions may exhibit reduced surface-water quality due to a higher presence of wastewater effluent or agricultural runoff. Conversely, flood waters may represent the poorest water-quality conditions because of high turbidity and contaminants mobilized by flood waters (e.g., pathogens and pesticides from agricultural watersheds).

Many large rivers in the United States (e.g., the Missouri, Ohio, Mississippi, Columbia, etc.) are regulated by locks and dams. Consequently the water-level fluctuations over the year (except during peak flood periods) are typically not extreme. However smaller streams in the western United States are typically characterized by more variable flow conditions with concentrated high-flow events in the winter and spring.

The resulting water management strategies for these different conditions reflect the basic hydrology of the specific river system. For example, in the Russian River, a relatively small river in California, the abovementioned inflatable dam is installed during low flow periods by the Sonoma County Water Agency to backup the water behind the dam so that the collector wells could produce enough water to meet summer water demand (Metge et al. 2010). Also, in regulated rivers the dams may have to release water during low flow periods to meet quality and quantity issues of the surface water.

5.8.2.4 Watershed Protection

RBF systems need source water protection to prevent degradation of water quality. River-water quality is a function of watershed activities. Both point and nonpoint sources of pollution could have adverse effects on river-water quality. As a result, land use activities in the watershed, locations of all point sources, and information on the contaminant types and quantities must be available to water utilities if the RBF system is located at a downstream location. Farm chemicals can be found in rivers in high concentrations during spring runoff or early summer periods (Ray et al. 2002c). Particularly, their concentrations can be high in smaller rivers.

Water utilities rarely have authority to regulate land use in watersheds unless it owns the land or the use of public land can be regulated by an agency that also regulates the water utility. If a RBF system must be placed in a river where the watershed land use is difficult to assess and control, the utility must make substantial efforts to monitor temporal changes in river and groundwater quality.

5.8.2.5 Post RBF Treatment

Water treatment needs after extraction from RBF wells can be significantly different depending on the intended use of the water and the local water quality regulations that must be met. As discussed in Sect. 5.1, it is critical to define the objectives of the RBF at the outset of the planning process. If RBF is used as pre-treatment for other treatment processes, it can smooth out variability in surface-water quality and help promote more efficient treatment by the secondary treatment process (Wang et al. 2002). Also, RBF can reduce DOC surface water, thus reducing the formation of trihalomethanes and other disinfection by-products. When treatment plants treat very clear water produced from RBF systems, certain precautions may need to be taken. In sedimentation tanks, light can penetrate deep into the tank and form algae. This was the case at a treatment facility at Louisville, Kentucky, which decided to cover parts of the basins to minimize algae growth.

5.8.3 Monitoring Parameters

Monitoring is a key component of system operations to ensure stable trends are maintained. Some key monitoring parameters are described below.

5.8.3.1 Water-Level Monitoring

Water levels and river stage data (preferably correlated to river flow) are basic, relatively inexpensive data to collect as part of an ongoing monitoring program to support RBF operations. A monitoring well network surrounding the RBF facilities with some wells preferably located on the opposite side of the river from the collection system (for smaller rivers) is important to gain an understanding of the hydraulic response of the aquifer to pumping stresses and surface water and groundwater interactions along the riverbed. Water levels in the production well, river, and monitoring well should be measured and referenced to a common data elevation in conjunction with recording corresponding pumping rates.

5.8.3.2 Water Quality

There is little cost required to routinely and frequently measure such water-quality parameters as alkalinity, conductivity, dissolved oxygen, particle counts, pH, temperature, and turbidity at several wells and river locations to generate a database monitoring general water-quality trends. Such monitoring can be incorporated into a larger regional-scale program as part of an early-warning detection system (see Sect. 5.8.4.3). Other water-quality parameters that are more expensive to sample and analyze may either be specified by permit requirements (e.g., metals, pathogens, pesticides, petroleum products, phenols, and volatile organic compounds) or may be monitored at the discretion of the operator on a more selective basis (e.g., pharmaceutical and personal care products).

Operators of production wells may easily measure the low-cost water-quality parameters noted above, including turbidity and particle counts, on daily or more frequent intervals using sensors and data-acquisition systems. Measurement of dissolved oxygen in production wells is less critical given the variability caused by turbulence and exposure of the water to the atmosphere. Monitoring frequency for regulated compounds is usually dictated by the regulatory agency. Non-regulated water-quality parameters of increasing concern may be monitored: (a) if it is desirable to establish a baseline trend, or (b) to address public concerns.

Monitoring wells are ideal for measuring trends in water-quality parameters between the river and the production well(s). Automated instrumentation such as probes with data loggers can be setup to monitor water levels as well as temperature, conductivity, pH, and oxygen. Other redox-sensitive parameters such as nitrate, ammonia, iron, and manganese may be monitored at desired frequency to assess possible significant changes in water quality. Monitoring wells should be designed to sample water from various depths so that the changes in water quality can be examined. Other parameters measured at the pumping well or river must also be measured at the monitoring wells.

Automated measurement of river stage and selected water-quality parameters such as temperature, pH, specific conductance, dissolved oxygen, and turbidity are possible using multi-probe sensors. Additionally, regulated parameters must be measured as required in the regulatory permit. If water quality is expected to change due to flooding, key water-quality parameters and river stage must be monitored until the flood passes through the site. Pollutions events (e.g., chemical spills) in navigable rivers can affect pumping-water quality. These chemicals are to be monitored both in surface water and groundwater as the plume passes through the RBF site.

5.8.4 Analysis of Operational and Monitoring Parameters

Analysis of monitoring data provides guidance to RBF system operators. There should be continuous feedback between data collection, data analysis, and system operations. The following describes common methods of data analysis that may provide useful insights into the operation of RBF systems.

5.8.4.1 Yield/Production

The production of the RBF over time should be monitored as part of a system-operations program. It is important to assess the system production under similar conditions to ensure an "apples to apples" comparison of data over time. It is particularly important that adequate system testing and monitoring occur during the initial startup phases so that benchmarks are established to compare future production data over time with this initial data set. This allows an assessment of whether gradual reductions of production may be occurring from increasing well loss (which may be mitigated with well redevelopment) or other factors such as aquifer compaction (which is not easily mitigated) and/or clogging of the riverbed.

The specific capacity of a pumping well is an easily measured parameter that can be tracked over time to evaluate changes in well efficiency. The specific capacity is defined as the pumping rate divided by the water-level drawdown in the well relative to static conditions. It is important that the drawdown used reflects stable water-level conditions. Higher specific-capacity values indicate more efficient well production. The specific capacity of an RBF facility can vary seasonally due to environmental factors (e.g., clogging of the riverbed or water temperature) or over longer timeframes with changes in the wells such as scaling or biofouling of the well screen. As with the cases noted above, it is important to measure specific capacity of a well during startup operations to establish an initial value (baseline) for later comparison.

Another method used to evaluate the hydraulic effect of pumping is to plot water levels versus time for the production and monitoring wells to evaluate seasonal and long-term trends. In addition, water-level contour maps created from water levels monitored at these wells can help evaluate the capture area of the production well, recharge areas, and groundwater flow paths. For example, a comprehensive contour map can delineate the cone of depression of a production well and the infiltration area from the river.

Numerical flow modeling calibrated to actual data can provide a means for determining proportions of bank filtrate/groundwater and travel times of bank filtrate, evaluating different operational scenarios, or predicting future trends. The most commonly used model for yield prediction is the USGS finite-difference model MODFLOW (see Harbaugh et al. 2000). In most instances the collector well is treated as a pumping center without regard to the flow through individual laterals. MODFLOW has been used in groundwater studies for several decades and it is easy to setup using graphical user interfaces such as the Groundwater Modeling System, Groundwater Vista, Processing MODFLOW, or Visual MODFLOW. In a limited number of cases, laterals have been explicitly treated in MODFLOW modeling. Ray et al. (2002c) considered laterals to be high-permeability zones where the hydraulic conductivity values were two-to-three orders of magnitude higher than those of the porous media. Schafer (2003) used a constant head differential between the river and the laterals. More recently Bakker et al. (2005) used analytic elements to simulate flow to collector wells in multi-layer aquifers. This approach allows the delineation of radials, internal friction losses in pipes, and incorporation of skin effects. Finite element models such as FEFLOW (DHI-WASY GmbH 2008) and SPRING (delta h Ingenieurgesellschaft, Witten) are used in many European countries as management tools for estimating flow to RBF systems.

In some instances unsaturated conditions develop under riverbeds (e.g., Louisville, Kentucky; Sonoma County Water Agency, California) when the rate of natural infiltration from the river in the vicinity of the well cannot cope with the production needs of the well. Under such conditions, pressure drop under the riverbed becomes excessive and air can enter from the land side to underneath the riverbed. Excess negative pressure may also cause the release of dissolved gases from water. The flow in such instances is three-dimensional involving both the water and air phases. Computer models such as TOUGH2 (Pruess 1991) or FEHM (Dash et al. 1997) can be used to simulate these conditions. These multi-phase and multi-dimensional models are difficult to use and require an enormous amount of parameters. Without relevant and adequate data, it is not possible to validate these models and thus the use of these models is limited to well-characterized sites. However, these models, if available and setup for a given site, can be used to conduct simulations for "what if" scenarios to examine the impact of an associated variable (a form of sensitivity analysis).

5.8.4.2 Water Quality Trend Monitoring

Analysis of water quality trends is an important component of an RBF operational program. The following describes how water-quality monitoring data can be analyzed to guide system operations.

Trend monitoring is essential for identifying changes in water quality in production and monitoring wells. This can be done by plotting water-quality parameters with time. Correlated parameters must be plotted together. This allows the utility manager to visualize possible water-quality trends during the operation of the RBF. Consistent changes over time in water quality in the monitoring well(s) between the river and the production well(s) indicate changes that can soon be expected in the production well(s). Trend monitoring is also important to evaluate the effects of flooding and/or pollution events in rivers and significant changes in operational practices.

5.8.4.3 Water Quality Modeling

Water-quality modeling at many RBF sites in the United States is generally addressed using the mass-transport model MT3D (Zheng 1992) because this model runs on the same grid used for running MODFLOW. Although MT3D is quite easy to use the model needs to be validated for the site and adequate transport parameters, such as degradation, dispersivity, porosity, and sorption, must be available for site-specific conditions. This model is of limited use if such site-specific data are not available.

The FEFLOW model commonly used in Europe may also be used for mass-transport simulations. This model has site-specific data requirements and limitations similar to those noted for MT3D. More recently, reactive transport models such as PHT3D (Prommer and Barry 2001) are used to examine the impact of geochemical reactions, organic-carbon biodegradation, and river-water temperature fluctuation on filtrate quality. PHT3D uses the same MODFLOW and MT3D grids and uses the database of PHREEQC (Parkhurst and Appelo 1999) for geochemical reactions. Although the model is somewhat complex it appears to have significant applications for RBF operations.

5.8.4.4 Emergency Early Warning Systems

Pollution events created by spills or releases of chemicals from barges, industries, municipalities, and ships in and along rivers can be expected, particularly for large navigable rivers. Early-warning systems are installed to determine and monitor accidental and/or illegal pollution events on the river. A network of monitoring stations along the river, combined with an automatic reporting system, is necessary to keep authorities and utilities informed of such events in a timely manner.

Online monitoring of several parameters, daily laboratory analysis of composite samples, and daily reference samples from each monitoring station should be sufficient to reliably detect significant pollution events. The following questions about the behavior of the pollution event have to be answered by the utility to fully utilize alert information:

- Time of arrival of the contaminant plume at the RBF site
- Concentration in the river water (as determined by a river flow and transport model)
- Biodegradability during bank filtration (e.g., test-filter experiments)
- Time of arrival in well water (as determined by a groundwater flow and transport model)
- Possible concentration in the well water

Monitoring wells located between the river and the production wells are necessary to validate the results and the precision of these tools (Fig. 5.13).

Fig. 5.13 A monitoring well located between the Rhine River and the wells

Information needs are the basis for the design of an early-warning system along a river. Basic questions to be answered in a case of a pollution event are:

- Time and site of spill
- Quantity and type of pollutants
- Actual hydrological data (river-water level, discharge)
- Effects on the ecosystem of the river (e.g., dying fish—demonstrating the extent of the event).

A superior early-warning system would also address the following questions. From the perspective of a waterworks manager, answers to these questions provide significant additional information:

- Time of arrival of the pollutants at site
- Range and duration of concentration in the river water at site
- Biodegradability of the pollutants during subsoil passage
- Transport-related physico-chemical data of pollutants (e.g., density, Henry coefficient, solubility, and sorption coefficient/K_{ow})
- The effect on well-water quality
- Time of arrival and concentration in well water
- The effect on drinking-water quality
- Any possible lasting effect on the biota at the sediment/water interface once the spill passes through the RBF site

If monitoring stations detect hazardous substances in river water at critical concentrations, messages must be immediately transmitted to one of the alert stations, maintained by a state agency, of the early-warning system. Depending on the results of a first data check and any available additional information the full reporting system may be initiated. If the identified pollution event is sufficiently extreme the notified alert station then transmits (using beepers, email, fax systems, telephones, text messages, and/or other electronic systems) an informational or warning message to all other alert stations, to the environmental and regulatory authorities, and to all potentially affected waterworks along the river.

5.8.4.5 Examples (Europe/USA)

Along significant rivers where accidental pollution is not a unique or unusual event but one that occurs repeatedly (e.g., the Rhine River with its accumulation of chemical industry), waterworks can significantly reduce the cost and effort of pollution management by having a good emergency-response plan. Such plans greatly facilitate rapid decision making regarding the adequate operation of wells and subsequent treatment facilities during severe pollution events. Good emergency-response plans incorporate communication with government authorities and media contacts to help provide information to water consumers during severe pollution events, thus helping to stabilize the situation.

To minimize the reoccurrence of severe pollution events, follow-up efforts addressing possible procedural changes at the site of the event and additional prevention measures have to be discussed, considered, and, where possible, implemented. Such follow-up review must include the individuals and/or institutions responsible or possibly responsible for the pollution event, the appropriate environmental and regulatory authorities, and any affected waterworks. Early-warning systems are intended to be learning systems, as well as response systems, and should be continuously improved to address changing situations.

Early-warning systems are operational around the world in numerous rivers serving as source waters to riparian populations (AWWARF 2001). These rivers include the Elbe in the Czech Republic and Germany, the Han in the Republic of Korea, the Llobregat in Spain, the North Saskatchewan in Canada, the Ohio in the United States (the Ohio River Sanitation Commission—ORSANCO), the Rhine (the International Commission for the Protection of the Rhine River—maintaining nine international stations and twenty national stations along the Rhine River through Switzerland, Germany, and The Netherlands), and the Seine in France. In case of pollution events monitoring stations located along these rivers report concentrations of chemicals so that the water utilities are warned to take appropriate actions.

5.8.5 *Maintenance*

Given the simplicity of RBF systems compared to conventional purification technologies the RBF systems generally require relatively low maintenance. Since much of the water treatment is accomplished through natural processes there are minimal mechanical and electrical components to the system. Enhancements to RBF systems (see Sect. 5.7.4) require additional maintenance activities—which can be significant if the enhancements have not been properly designed for site conditions.

In general, however, the maintenance of RBF facilities consists of maintaining the integrity of the screened well or lateral intake area from scaling or biofouling, maintaining the pumping equipment, and maintaining any appurtenances (e.g., disinfection-process equipment). Under favorable water-quality conditions (e.g., aerobic conditions with low DOC and microbial concentrations), RBF facilities may not need to have their well or intake lateral screens redeveloped for decades.

Well maintenance—The efficiency of a well in producing filtered water should be at a maximum (assuming adequate development) immediately following the installation and development of the well. After the initial installation the efficiency of the well will decline as the well is pumped. The rate of decline will be a function of:

- Well design (including drilling and installation methods)
- Pumping rate and schedule
- Aquifer character
- Water quality

It is generally considered prudent to perform screen maintenance after the specific capacity has declined 20–50% from the original specific capacity. If the maintenance is conducted too late its effectiveness may be reduced as permanent loss of capacity may have resulted from compaction, precipitation, or bacterial plugging (which cannot be removed or rehabilitated using generally accepted practices).

Screen maintenance processes involve physical and chemical methods resulting in the removal of the plugging (bacterial and inorganic) in and around the well screen and gravel pack. General procedures involved with maintenance include:

- Assessment of the loss of capacity (development of maintenance plan)
- Removal and/or evaluation of pumping equipment
- Redevelopment
- Disinfection
- Post-maintenance testing

Maintenance procedures will be different if bacterial plugging is suspected as being the primary problem from what it would be if the plugging was believed caused by the precipitation of carbonates or the migration of fine particles. If chemicals are used, their compatibility with the well materials, the aquifer, and the environment should be fully evaluated. Screen maintenance generally involves the use of both chemical and physical methods in conjunction.

Chemical methods include:

- Biocides
- Carbon dioxide
- Acids
- Dispersants and/or surfactants

Physical methods include:

- Surging (double disc and surge block)
- Air-lifting
- High-pressure or high-volume jetting
- Low-pressure filter purging and/or washing
- Blasting (using explosives and/or impact generators)

It may be necessary to conduct short pumping tests during development to determine if more treatments or different methods are warranted. Based on its maintenance history, with time it may become cost effective to re-drill or replace a well.

Collector well maintenance—As with any water well, a collector well will require maintenance to restore lost well efficiency and capacity. With the data collected and graphed as part of the monitoring program, declines in well performance can be observed and maintenance can be anticipated and scheduled for opportune times to minimize disruption to normal service. Sonoma County Water Agency collector well 1 showed a specific capacity of 0.1893 m^3s/m of drawdown in 1959. In 2003 the well was rehabilitated and after cleaning, the specific capacity was found to be 0.0591 m^3s/m. In 2008 the well was cleaned again and the resulting specific capacity was 0.0696 m^3s/m. For collector well 2 the specific capacity was 0.1798 m^3s/m in 1959 and after a recent (2008) rehabilitation effort it could achieve a maximum specific capacity of 0.1086 m^3s/m. Two long laterals of a new well (collector well 6) had a specific capacity of 0.2648 m^3s/m in 2006 (soon after installation) but had been reduced to 0.2205 m^3s/m by 2008.

Typical collector well maintenance includes the following:

- Standard well inspections every 5–7 years
- Normal pump inspection and periodic maintenance
- Screen cleaning and redevelopment
- Replacement of lateral well screens

It is also recommended that an in-well underwater inspection be made of the collector well under operating conditions first after the well has been in operation for about seven years and then once every five years thereafter. During such inspections, observations of the physical conditions in the well, well-screen conditions in the sections of the laterals closest to the caisson of the collector well, and flow measurements taken from the laterals are made by a diver.

Maintenance is most effectively accomplished by specialized equipment inserted into the individual laterals while the central caisson is kept dewatered. This equipment flushes out any loose scale, encrustation, and or fine sands that are loosened during the cleaning process. This cleaning and redevelopment program removes debris from inside the well screen, accumulations within the well-screen slot openings, and deposits from outside the well screen in the aquifer and gravel-pack materials.

General methods that can be used to clean/redevelop lateral screens include:

Chemical Methods

- Biocides (Cl_2)
- Acids (muriactic, sulfamic, acetic)

Physical Methods

- Surging/flushing with or without packers
- High-pressure jetting
- Impulse generators

If a well is the only source of water supply for a water utility, well-screen cleaning and redevelopment may also be performed while the collector well remains in service by using specialized maintenance procedures. With these methods redevelopment and cleaning can take place while the collector well continues uninterrupted pumping into the system. This illustrates one aspect of the flexibility of collector wells.

Maintenance can also include installation of new and/or additional lateral well screens when older screens (usually constructed of mild steel) have corroded or deteriorated with time or become excessively plugged. Many of these older collector wells with the mild steel screens require new lateral screens after forty or more years of operation. However, most of the collector wells built in the last twenty years have used newer stainless steel well screens that should last longer and be more resistant to normal corrosion.

Typically more ports are built into the bottom of a caisson than there are active laterals. New laterals can be installed into an existing facility by installing new port assemblies and projecting new lateral screens. Previously unused ports are used to project screens into the aquifer. This type of maintenance can restore well capacity where older screens need replacement and/or supplement existing well screens to develop additional capacity (where available) or to extend well life.

5.9 How Well Does It Work?

RBF systems improve the physical, chemical, and microbial quality of pumped water.

Physical parameters that get improved by RBF are suspended solids (often measured as particles of different diameters), temperature, and turbidity. Temperature of groundwater does not fluctuate significantly compared to surface water. In temperate climates, the seasonal variation in surface water temperature can range from near 0 to 36°C (Schoenheinz and Grischek 2010). This variation in temperature not only affects infiltration rate of water, but also affects the chemical and biological reactions occurring in water.

Figures 5.14 and 5.15 show the counts of 2 and 5 μm particles per milliliter of water in the Russian River (California, USA) and in its bank filtrate. These sizes bracket the size of *Cryptosporidium parvuum* oocyst sizes found in fresh water systems. For both sizes the removal efficiency of collector well 1 of the Sonoma County Water Agency is more than 2 logs. The lateral screens for this well are located about 15 m below the riverbed.

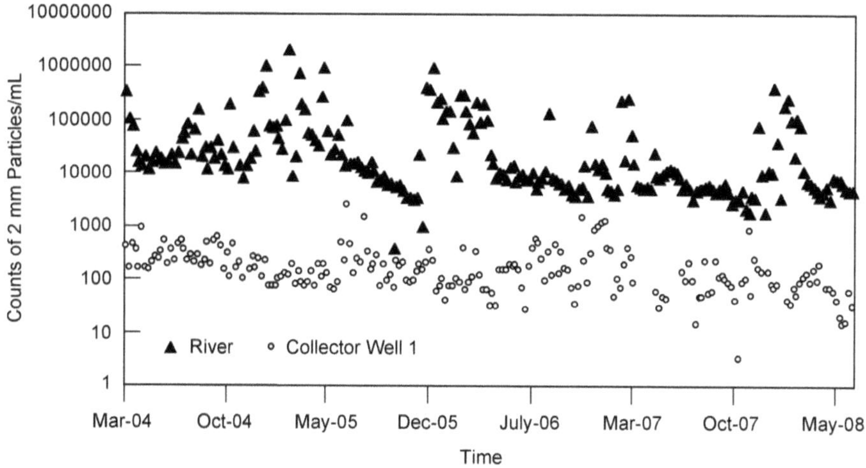

Fig. 5.14 Counts of 2 μm particles per ml of water in the river and in the filtrate from collector well 1 of Sonoma County Water Agency

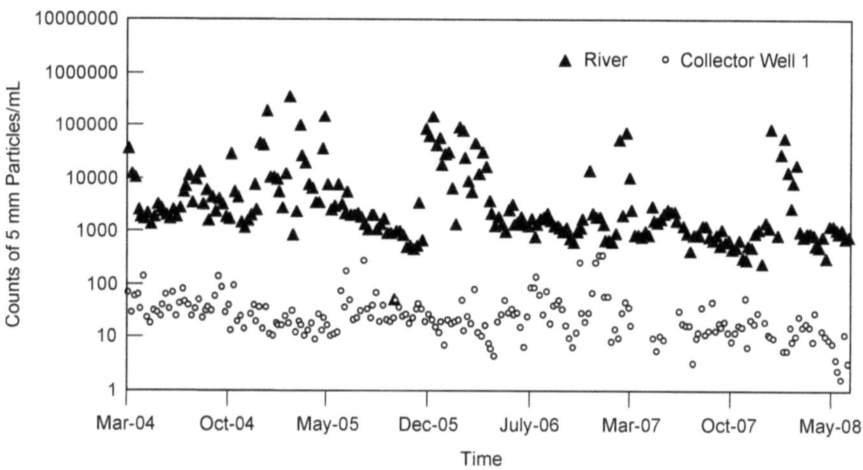

Fig. 5.15 Counts of 5 μm particles per ml of water in the river and in the filtrate from collector well 1 of Sonoma County Water Agency

Seasonal temperature of the water in the Russian River varies between 10 and 25°C. The temperature variation for collector well 1 is, however, just slightly over half that, with the temperature varying only 8°C, between 15 and 23°C (see Fig. 5.16). For the RBF site at Louisville, Kentucky, the temperature of the river water varies between 0 and >30°C but the filtrate temperature only varies between 12 and 25°C (Wang 2002).

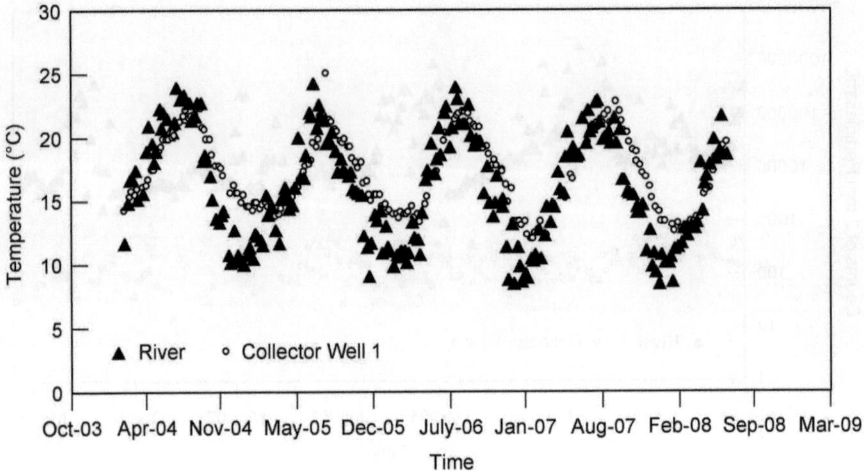

Fig. 5.16 Seasonal variations in river water temperature and its effect on the filtrate from collector well 1 of Sonoma County Water Agency

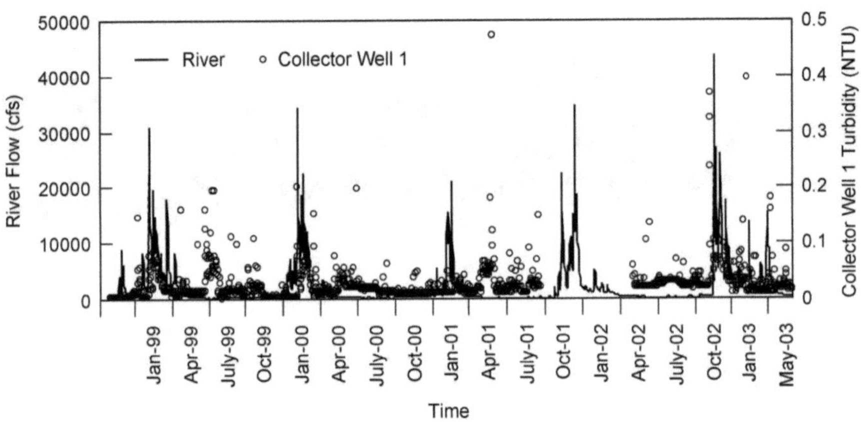

Fig. 5.17 Variations in flow of river water and filtrate from collector well 1 of Sonoma County Water Agency. The scale for the turbidity on the *left* Y-axis is in log scale

Turbidity of river water is closely associated with its flow. Figures 5.17 and 5.18 show the variation in river-water flow and river-water turbidity and the relatively stable associated turbidity observed from collector well 1 of the Sonoma County Water Agency. From Fig. 5.17 it is evident that after a flow peak in the river a small peak of turbidity sometimes appears in collector well 1. However the turbidity of the well water stays below 0.5 Nephelometric Turbidity Units (NTU). Figure 5.18 records an up-to 3 log reduction in turbidity.

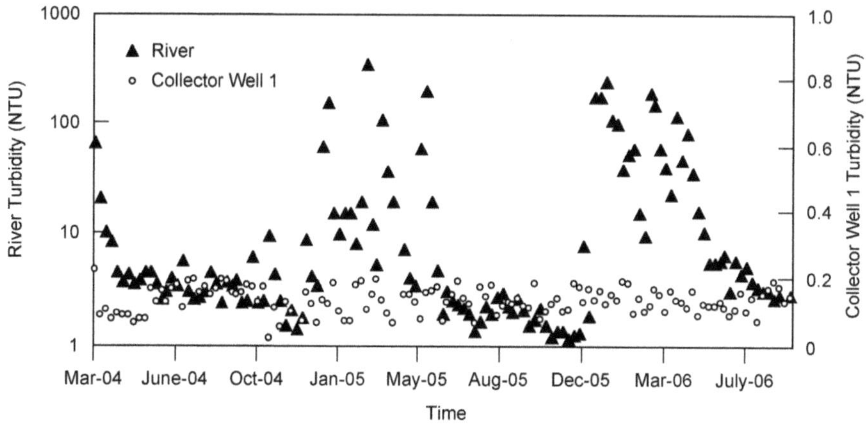

Fig. 5.18 Variations in turbidity of river water and filtrate from collector well 1 of Sonoma County Water Agency. The scale for the turbidity on the *left* y-axis is in log scale

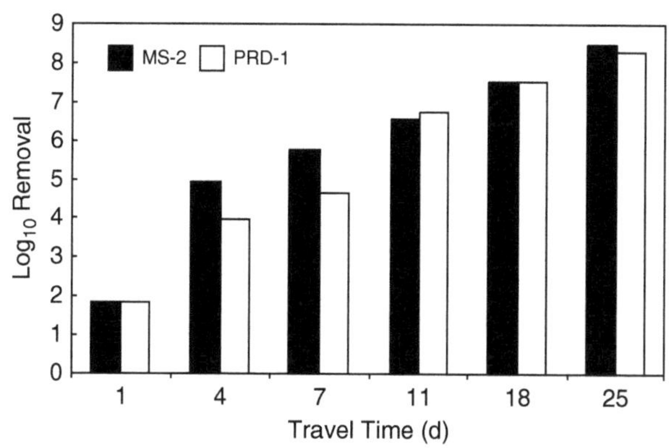

Fig. 5.19 Removal of bacteriophages MS-2 and PRD-1 as a function of travel time through dune sand. (Replotted from data of Schijven et al. 1999)

Removal of microbial pathogens or their indicators is also dependent on travel time or flow time of water. Pumping rates and well type affect travel time. Schijven et al. (1999) studied virus removal using indicator viruses MS-2 and PRD-1 during dune-sand infiltration of Rhine River water. As is evident from Fig. 5.19, within 24 h nearly 2 log removal is achieved. Within 25 days approximately 8 log removal is achieved. Medema et al. (2000) examined the removal of somatic and F-RNA coliphages, sulfite-reducing spores, and thermo-tolerant coliform at a Meuse River RBF site (Fig. 5.20). Over 30 days removals varied between 4.5 and 6.5 logs depending on the pathogen indicator. These two studies show significant removal with time. Both sites used vertical wells and as a result the travel times for the water between the river and the wells were slow.

Wang (2002) examined the removal of aerobic spores, pollen, and diatoms during a one-year study at the RBF site of Louisville, Kentucky after the collector well

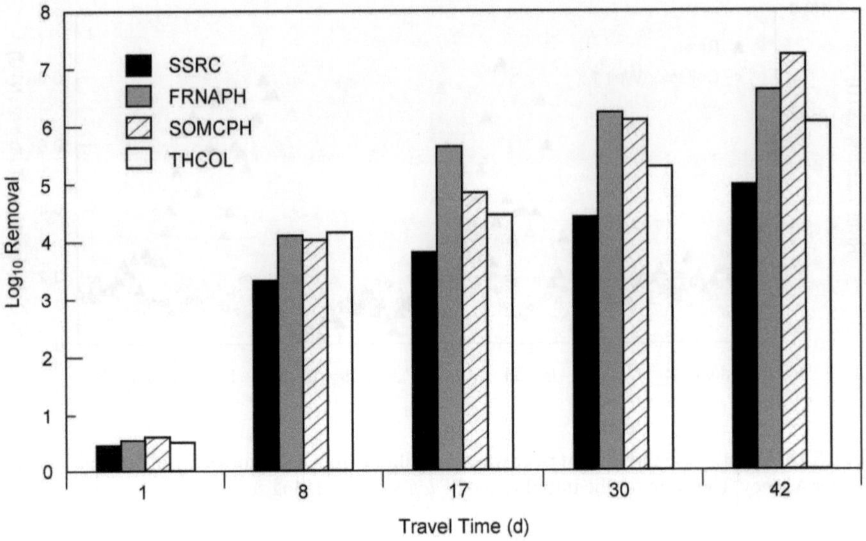

Fig. 5.20 Removal of thermo-tolerant coliforms (THCOL), spores of sulphite-reducing (SSRC), somatic coliphages (SOMCPH) and F-specific RNA bacteriophages (FRNAPH) by RBF. (Replotted from data of Medema et al. 2000)

was installed. The aerobic spore counts in the river varied from several hundreds per 100 l to nearly 1,000,000 per 100 l. However, in the filtrate from one of the laterals (lateral 2) the maximum aerobic spore count was 3. Most of the time, aerobic spores were not detectable in the collector well water. A similar trend was observed for pollen. For diatoms, almost all were removed.

The removal of TOC is important from the perspective of the efficient operation of activated-carbon filters and for the reduced formation of disinfection byproducts. At collector well 6 of the Sonoma County Water Agency more than 50% removal of TOC is observed (Fig. 5.21).

From 1977 to 1997 improved pollution mitigation reduced the concentration of DOC in the Rhine River at Düsseldorf, Germany, from about 5 to about 3 mg/l (Fig. 5.22). The DOC levels in the RBF wells there dropped correspondingly from 2.5 to about 1.5 mg/l.

It is clearly evident that, overall, natural filtration processes used at RBF sites significantly improve the biological, chemical, and physical quality of the extracted drinking water. Water utilities employ various strategies regarding selecting sites for RBF systems where riverbeds are less likely to be clogged by sediments and/or biological solids, well placement relative to the rivers, the use of appropriate pumping rates, and (primarily in navigable rivers) establishing networks of upstream sensors to report pollution events. Monitoring wells providing information on subsurface processes and for early detection and warning of possible problems are another key element in the successful operation of RBF systems. Together all of these efforts lead to the enhanced sustainability of RBF systems.

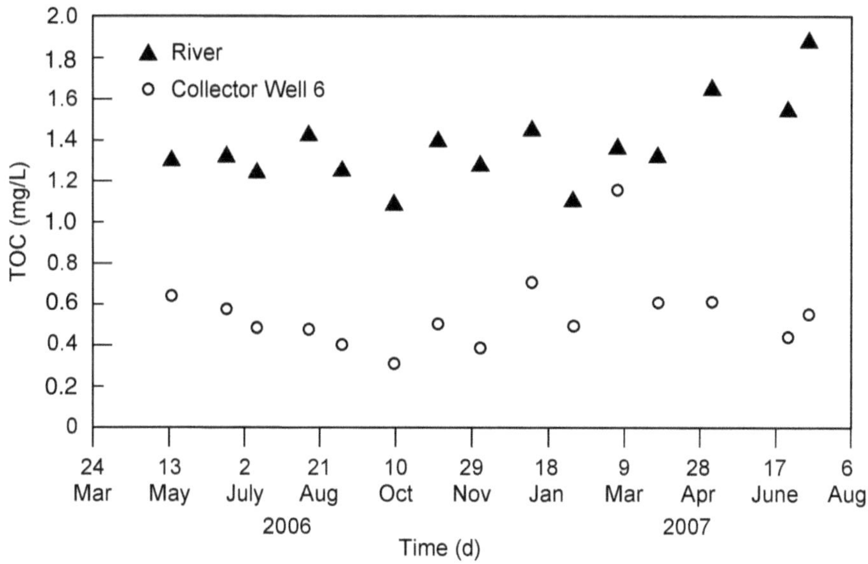

Fig. 5.21 Variations in total organic carbon content of river water in filtrate from collector well 6 of Sonoma County Water Agency

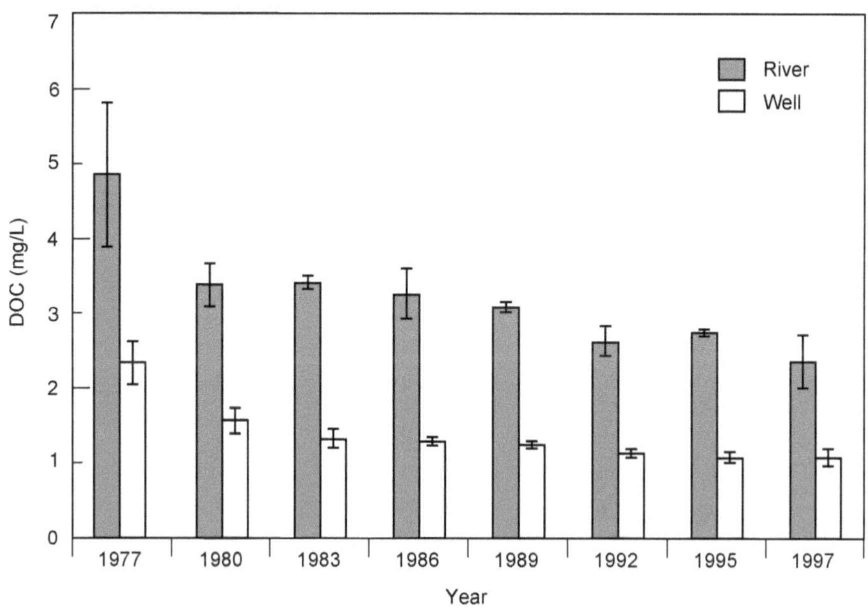

Fig. 5.22 DOC concentration in the Rhine River at 731.5 km and in the well. (Plotted with data from Schubert 2002)

5.10 Performance Assessment of RBF Systems

The key aspects determining filtrate quality are (a) source-water quality, (b) hydraulic connection between the river and the aquifer where the well screens are located, (c) pumping rate, and (d) quality of aquifer solids and groundwater. Of these, source-water quality is the key driver in determining filtrate quality. Any filtration system, engineered or natural, is heavily affected by the quality of the feed water.

For example, in the European reconstruction period following World War II many rivers (e.g., the Elbe and Rhine) were heavily polluted. Industrial and municipal wastes with little or no treatment were discharged into the rivers. Consequentially the quality of the filtrates developed from such rivers was low. Breakthroughs (at pumping wells) of many industrial organic chemicals were common (Schmidt et al. 2003).

Post-World War II industry along the Upper Elbe River valley discharged a wide range of organic contaminants into the river. Mixed with minimally treated or untreated urban sewage, the DOC comprised a complex mixture of both easily degradable and refractory substances. In the 1980s cellulose-processing plants, paper mills, and the pharmaceutical industry were significant contributors of industrial effluents. From 1988 to 1990 the average DOC concentration on the left bank of the Elbe River at Dresden-Tolkewitz was 24 mg/l and the UV-absorbance (at a wavelength of 254 nm; UVA254) was 55 m^{-1}.

An RBF system using a flow path of approximately 100 m along a cross-section of the Elbe River at Dresden-Tolkewitz was able to reduce the DOC concentration in filtrate to about 20% of the source-water input concentration (Nestler et al. 1991). Given the high load of organic pollutants in the source water, problems still remained with the odor and taste of the bank filtrate and with the formation of disinfection byproducts. Figure 5.23 gives an impression of the organic load in the

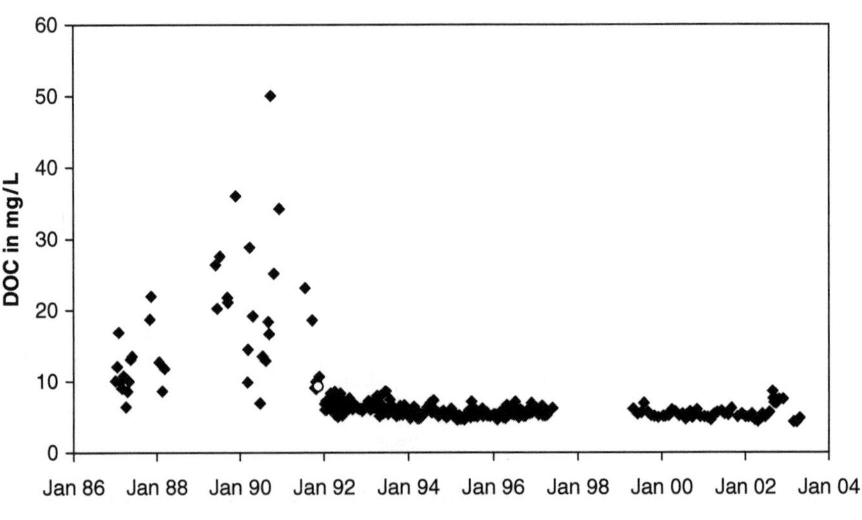

Fig. 5.23 DOC concentration (mg/l) in Elbe River water 1986–2003. (Grischek 2003)

Elbe River from 1987 to 1992. The small data set does not allow the calculation of mean concentrations and the variation in the data is very high. Furthermore, the sampling point on the left bank of the Elbe River at Dresden-Tolkewitz was affected by wastewater outflows only a few kilometers upstream.

Results from seventeen measurements in 1991–1992 at a cross section at Dresden-Tolkewitz showed a mean DOC concentration of 6.9 mg/l in the Elbe River and 3.4 mg/l at a monitoring well near a production well. From these results a reduction of DOC concentration of a little more than 50% can be seen as an effect of the RBF process. An extreme flood in August 2002 caused a significant increase in the DOC concentration in the river water but no increase in the DOC concentration in the bank filtrate. Results from seven measurements in 2003 at the same cross section showed a mean DOC concentration of 5.6 mg/l in the Elbe River and 3.2 mg/l at the same monitoring well sampled in 1991–1992. As a result of river filtrate mixing with groundwater the mean DOC concentration in the raw water from all wells in this RBF system was found to be 2.6 mg/l. These results demonstrate that even the sustained period of heavy pollution of the Elbe River did not limit further use of the RBF site.

Earlier at the Dresden-Tolkewitz RBF site a significant decrease in groundwater levels observed between 1914 and 1930 had been attributed to riverbed clogging generated by infiltration rates that had significantly increased after 1901 and the concentration of suspended materials in the river. In the 1980s the above-noted industrial effluents generated by cellulose-processing plants, paper mills, and the pharmaceutical industry in conjunction with high water abstraction caused unsaturated conditions beneath the Elbe riverbed, especially at the Dresden-Tolkewitz site.

Investigations of the riverbed using a dive-chamber, however, showed that the material responsible for the pore clogging in the gravel bed consisted of up to 90% inorganic materials (Heeger 1987). Heeger (1987) defined a leakage coefficient as the ratio of the hydraulic conductivity (in m s^{-1}) and the thickness of a clogging layer (in m). He calculated a leakage coefficient of about $1 \times 10^{-4}\ s^{-1}$ for the riverbed without bank filtration and a mean value of $5 \times 10^{-7}\ s^{-1}$ at RBF sites in and around Dresden.

The long-term process of riverbed clogging includes a series of building and destruction phases, which overlay a mean value. During floods generating sufficient hydraulic-transport energy the riverbed is eroded and the hydraulic conductivity of the riverbed is subsequently increased. Similarly long low-flow periods result in decreasing leakage values.

After the improvement of river-water quality seen in the Elbe River in 1989–1993, the hydraulic conductivity of the riverbed increased. In 1992 water levels similar to those seen in 1930 were observed. In 2003 groundwater flow modeling was used to re-analyze former assumptions regarding the clogging of the riverbed and groundwater flow to the production wells. From model calibration a reliable leakage coefficient of $1 \times 10^{-5}\ s^{-1}$ was determined. Water-level measurements in 2004 at low-flow conditions also indicated a slight decrease in clogging.

Looking at the long-term operation of the waterworks, it is obvious that observed clogging of the riverbed did not result in the closure of wells under the given conditions of an erosive river. During the 1980s additional organic pollution (e.g., from pulp and paper mills) caused an organic rich outer clogging of the riverbed. After a significant reduction of the organic load due to closure of industries (Fig. 5.23), a recovery of the hydraulic conductivity of the riverbed was observed (Grischek et al. 2009).

Perhaps a better case, today's Rhine is much cleaner than the river of the 1960s and 1970s. RBF systems along the Rhine no longer experience anoxic conditions and, except in the now far-less-frequent pollution events, breakthroughs of industrial organic chemicals are much less common.

Hydraulic connection between the river and the aquifer is the key to both water production and the water quality of the filtrate. If the river bottom is clogged with low-permeability materials then the water production is reduced. For river systems that do not experience riverbed scour conditions this situation becomes worse in the winter when the viscosity of river water is much higher than during the warmer summer season. If the pumping demand remains the same then additional drawdown would be needed to get the same amount of water. This can cause negative pressures below the riverbed and eventually the entry of air from landside to beneath the riverbed (Su et al. 2007). A clogged riverbed also increases the travel time of water from the river to the well and biogeochemical reactions affecting the quality of water. A redox sequence may begin where the oxygen from the source water is initially consumed by microbes and through abiotic reactions.

The aquifer travel time for water strongly influences water quality of RBF facilities. Pumping rates strongly influence the aquifer travel time for water and have a corresponding influence on water quality. In addition to pumping rates, the placement of wells with respect to surface water also affects water quality. For example, collector wells produce large amounts of water by placing the wells close to the river and installing the lateral screens near or directly beneath the riverbed. The resulting flow time of water from the river to the wells is then shorter than for typical vertical well fields placed farther away from the river. If all other factors are equal (e.g., surface-water quality, soil chemistry, etc.), the water quality for systems with reduced aquifer travel times will be lower than for systems with longer travel times.

Wells situated farther away from the river will have a relative contribution of surface water that will be lower than that of wells placed closer to the bank. Groundwater quality will more significantly dictate the quality of filtrate for wells situated farther away from the river. In developing countries where the groundwater quality is not fully assessed the introduction of large quantities of groundwater to filtrate may be risky in terms of pollutant entry to the wells.

Laterals lie in shallow depths under the riverbeds for very shallow collector wells or infiltration galleries. Placement of laterals at shallow depths beneath riverbeds can have implications during high-flow events for rivers that experience scouring. In such situations breakthroughs of chemicals, pathogens, and turbidity are possible.

Clogging of riverbeds can be a cause of concern for utilities that utilize collector wells for water production. For example the Sonoma County Water Agency in California and Louisville Water Company in Kentucky have both experienced compaction of riverbeds and resulting reductions in infiltration in the vicinity of their collector wells. Unsaturated conditions may also develop beneath the riverbeds causing a lowering of infiltration and a reduction in pumpage.

In the Tancheon River in the Republic of Korea a collector well is present just downstream from the discharge point of a wastewater treatment plant. High organic loads have caused the evolution of anoxic conditions reducing the nitrates to ammonia. Strong redox conditions also seem to cause the reduction of sulfates at this location.

Lincoln, Nebraska, in the United States, has two large-capacity collector wells situated on an island in the Platte River. The Platte River flows through a predominantly agricultural area where the spring runoff water contains large amounts of pesticides. As the dilution with groundwater is minimum, these wells experience breakthroughs of certain pesticides, which must be subsequently removed through advanced treatments including ozonation and activated carbon.

Proper selection of RBF sites not prone to clogging is ideal to avoid problems with water quality. If the river hydrodynamics allows for particle transport then the riverbed particles that are removed by scouring may be replenished by material migrating from upstream. Placement of wells on the inside of a bend in a river can produce more water than placement of wells on the outside of a bend in a river. Smaller collector wells or wells installed some distance away from a river will reduce clogging of the riverbed caused by aquifer compaction.

At the Sonoma County Water Agency one of the collector wells experiences enhanced turbidity when the river flow is high. This occurs most commonly during winter months when the water demand is low. In these circumstances this well is taken out of service once the river stage exceeds a threshold flow value.

Cedar Rapids, Iowa, is a medium-size city with four collector wells and several dozen vertical wells. The collector wells, although having relatively short travel times, still remove a fraction of nitrates and pesticides typically present in the river water. Nearly every year high levels of spring-runoff water bring large amounts of pesticides into the RBF system. During such seasonal events the Cedar Rapids collector wells currently tend to capture pesticides in their filtrate. Consequently the city is considering installing a new collector well some distance away from the river with laterals that will not lie directly beneath the riverbed.

5.11 Areas of Future Study and Technology Development

5.11.1 Mechanisms of Natural Filtration

Significant research is currently focused on improving our understanding of the biological, chemical, and physical processes that form the basis for natural filtration. A better understanding of these processes will result in improved assessments regarding the reliability of RBF and enhance our ability to address problems such as the removal of pathogens and trace organics. One problem in particular, removal and/or inactivation of *Cryptosporidium* oocysts, continues to challenge RBF operations as this pathogen is resistant to disinfection by chlorination (Macler and Merkle 2000). Research is consequently being conducted to evaluate key factors that may strongly influence the ability of alluvial deposits to remove or attenuate the migration of *Cryptosporidium*. These factors include the composition and concentration of DOC, grain-size distribution(s), and grain-surface chemistry.

The presence of metal oxides (e.g., predominantly aluminum and iron oxides) on surfaces of soil grains may influence the ability of such granular material to remove pathogens. Abudalo et al. (2005) showed that *Cryptosporidium* migration was attenuated in iron oxide coated quartz sand. The results of column testing have shown that alluvial material that exhibit higher concentrations of metal oxides on grain surfaces are more effective in attenuating the migration of pathogens such as *Cryptosporidium* oocysts than granular material with lower concentrations of metal oxides (Metge et al. 2010). Metal oxide coatings on sediments appear to counter the secondary energy barrier caused by the repulsive forces between negative surface charges of soil grains and *Cryptosporidium*. Hence, sediment grains with metal oxides in sufficient concentrations can form an attractive force and promote pathogen attenuation (Metge et al. 2010). Recent investigations also suggest that the distribution of metal oxides in an uneven or “patchy” pattern on soil grains may increase the heterogeneity of a granular energy barrier, further enhancing the attraction or “stickiness” of the grains for pathogens (Elimelech et al. 2000).

Another environmental factor seeming to affect the ability of granular material to remove pathogens and other impurities from water is the concentration and type of dissolved organic matter, typically measured as DOC. Column studies with core material taken from depths near operational bank filtration wells indicate that increased concentrations of biogenic DOC decreased the ability of granular material to attenuate *Cryptosporidium* (Metge et al. 2010). Results suggest that DOC reacts with the metal oxides on the soil grains, tying up these sites so that possible pathogen-attractive sites are not as effective in attenuating pathogens as they would be under conditions of lower DOC concentrations. A related topic of research is whether anthropogenic sources of DOC (e.g., wastewater) are more detrimental in terms of pathogen removal/attenuation than natural sources of DOC (e.g., humic material from plants). RBF has been found to significantly reduce concentrations of trace organic compounds along with DOC (Hoppe-Jones et al. 2010).

Further studies in the use of surrogate organisms as indicators of pathogen transport are needed to determine the best monitoring tools for the most representative sampling of bank filtrate producing the least perturbation in the RBF system (Baveye et al. 2003) and to better predict pathogen breakthrough during floods.

Different clogging processes must also be studied in greater detail, especially the entrapment of gas bubbles in river and lake beds and their removal and the effect of small faunal organisms in increasing infiltration rates and limiting biofilm clogging.

5.11.2 Surface Water and Groundwater Interactions

The characterization of surface water and groundwater interactions is a complex endeavor and consequently is an active area of applied research. As discussed in Sect. 5.6, methods using temperature and surface geophysical methods (e.g., electrical resistivity and streaming potential) show promise as cost-effective approaches capable of characterizing the exchange along the riverbed between surface and groundwater. While current methods have been successful in addressing fine and medium sediments, methods have yet to be developed to determine infiltration rates and zones in coarse sediments, armored riverbeds, and large rivers with greater depth and high flow velocities.

5.11.3 Evolving Facility Design Considerations

With advances in drilling and micro-tunneling technologies, design enhancements and new configurations are being considered for RBF systems. These alternative approaches are being driven by regulatory, technical, and even political considerations requiring greater flexibility regarding where, within the aquifer, water is extracted relative to where surface pumping and support infrastructure are located. For example, increasing regulatory requirements (e.g., within the United States, the Endangered Species Act) often prohibit construction of RBF facilities within riparian habitat corridors supporting both terrestrial and aquatic species. Other reasons for evaluating new design approaches include reducing the aesthetic impacts of RBF systems along rivers, reducing the energy and maintenance requirements of RBF systems, and reducing the impact of flooding on pumping by RBF facilities by locating such facilities farther from rivers. New designs are also being developed to distribute pumping stresses over larger areas of the aquifer to achieve more efficient pumping (i.e., reduce well losses) in the case of collector wells—which have typically generated highly localized stresses on aquifers—while still realizing the benefits of centralized pumping facilities. Examples of recent design enhancements using horizontal drilling and micro-tunneling technologies are described below.

5.11.3.1 Modified Collector Well Design Using Horizontal Drilling Technology

The Sonoma County Water Agency in California recently constructed a horizontal collector well with a modified design that incorporated the use of horizontal drilling technology to install intake laterals that are significantly larger in diameter and longer than those previously achieved by conventional lateral installation methods. These new 45 cm diameter laterals extend up to 114 m in length compared to the conventional 30 cm diameter laterals that extend from 21.3 to 51.8 m into the aquifer. This approach helped overcome challenges resulting from the necessity of locating the collector well over 100 m from the river for aesthetic reasons and to address environmental regulations. Numerical modeling predicted that the enhanced design would increase production capacity and available drawdown, or well efficiency, by distributing the pumping stress over a larger area. Operational performance data has demonstrated that the enhanced design has greatly added to the performance of the collector well.

5.11.3.2 Use of Horizontal Wells in RBF Design

Des Moines Water Works provides water to more than four hundred thousand people in central Iowa. It operates six collector wells and a horizontal directionally-drilled (HDD) well located at the Maffit Treatment Plant 10 km west of the city of Des Moines. The caissons are about 12 m deep with a diameter of 4 m. Most of them have three laterals each 0.25 m in diameter with length ranging from 40 to 70 m. The HDD is 375 m long with a diameter of 0.3 m—however it provides only 10% of the total water produced at the site (Christopher Jones, Des Moines Water Works, personal communications 2009). An infiltration gallery (IG) at its Fleur Drive treatment plant pumps approximately 660,000 m^3/s of bank filtered water from the banks of the Raccoon River. This infiltration gallery is made of pipe segments spanning a 5 km length. The gallery is constructed from 1.5 m diameter pipes having 0.7 m-long sections with 0.2 m gaps between the pipes. The pipes are 10 m below the ground surface and 15–20 m laterally from the river. The Fleur Drive treatment works output averages about 2 m^3/s. The infiltration gallery was completed in 1931 and uses gravity feed from the gallery to the treatment plant. The Iowa Department of Natural Resources considers these wells (including the HDD and the IG) to be under the direct influence of surface water. Water from all HDD, IG, and RBF sources is treated at the two surface-water treatment plants operated by the city. Both the IG and RBF wells function well in terms of removing contaminants and pathogen indicators. However the HDD well appears to be less than successful in terms of water production.

5.11.3.3 Use of Microtunneling in RBF System Design

In its first phase of its conversion from surface water to bank filtrate the Louisville Water Company in Kentucky installed a collector well of 0.66 m^3/s capacity in 2000. The caisson of the well is 6 m in diameter and 34 m below ground level. The well has seven laterals with lengths from 61 to 73 m lying at 15 m below the riverbed.

The second phase of the Louisville Water Company conversion required a completely underground construction for the RBF system because of (a) aesthetic reasons and (b) lack of access to land along the riverbank. Four collector wells, spaced 1,800 m apart, were constructed with a capacity of 0.66 m^3/s each for winter pumpage with all four wells connected by a tunnel. The caissons of each well are 4 m in diameter and the laterals are located 70–74 m from the grade. The wells were connected by an open vertical shaft 1.2 m in diameter to the tunnel at 57 m below grade. The drop shaft is 12 m from the bottom of the caisson to the top of the tunnel. The tunnel has a diameter of 3 m and is 2,300 m long and runs parallel to the river while connecting the four wells and then curves toward the treatment plant. At the end of the tunnel is a 12.5 m diameter 38 m deep sump. The bottom of the sump connects to the tunnel through an 8 m diameter 20.5 m deep lined shaft. Four turbines pump the water to the B.E. Payne treatment plant.

To construct this facility initially the access shaft at the pump station was sunk to the desired depth. A tunnel-boring machine was then assembled at the bottom of the shaft and tunnel boring on the desired alignments began. The facility is projected to be completed in the summer of 2010 with the total cost of the project projected to be $ 36 million.

5.11.3.4 Optimization of Local Low-Cost Systems in Developing Countries

There is a high potential for the use of natural filtration techniques in developing countries where water-treatment budgets often cannot keep up with the population growth. Even if bank filtration is used only for pre-treatment the resulting water quality, especially concerning pathogen removal, is much better relative to direct surface-water abstraction.

Local systems should be optimized to gain the maximum sustainable pathogen removal. Simple maintenance procedures and agreements on well-head protection areas will significantly improve the water quality. Basic information concerning the advantages of certain distances between the well or drain pipe and the riverbed or bank and the importance of filter media can be simply provided to local users.

A recent example is the improvement of so-called Koop-wells in India. Filter sand and a piece of geotextile are now being used during well construction to increase the travel time of the water infiltrating from the riverbed from 5 min to more than 4 h (Sandhu et al. 2010).

5.11.4 Interaction of Aquifer and Intake Lateral Hydraulics

While many new designs for RBF facilities focus on distributing pumping stresses over larger areas of the aquifer by increasing the length and size of intake structures, a better understanding of how the aquifer and intake (pipeline) hydraulics interact is still needed. More research, in terms of field-data collection methods and modeling techniques, is needed to evaluate the potentially competing influences of aquifer hydraulics and pipeline hydraulics.

For example, longer horizontal intake laterals may result in more efficient groundwater hydraulics surrounding the intake lateral—but the head losses within the intake lateral created by the lengthened intakes may minimize or counteract this benefit, resulting in increasingly localized water production near the pumping center.

Recent modeling approaches using the "analytic element theory" (Bakker et al. 2005) are intended to help evaluate the interaction of groundwater and pipeline hydraulics so that an assessment of potentially diminishing returns can be made for various design configurations. To increase the usefulness of these models, however, field measurements of flow and velocity profiles along the length of the intake structures are needed. A recent study conducted by the Sonoma County Water Agency evaluated the flow profile along the length of intake laterals for three collector wells (including the long laterals described in Sect. 5.11.3.1; Environmental Resource Management 2010). In general the study found that most of the flow contribution came from the inner and outer portion of the laterals while the middle of the laterals contributed a relatively smaller portion of the total flow.

Better understanding of the interactions of groundwater and pipeline hydraulics may lead to new approaches for the design of pumping systems. Instead of centralizing pumping equipment, pumps may be distributed along the longer intake structures. A different alternative design, described below, uses a passive siphon-gravity system requiring little energy to distribute low flows and consequently exhibits a lower head loss within the intake laterals.

5.11.5 Sustainable Pumping Systems (Siphon and Gravity System)

In mountainous regions infiltration galleries are used as simple systems to abstract water from beneath the riverbed or along a riverbank. Water flows from the riverbed or riverbank to the point of abstraction under gravity so there is no need for any pump. Examples can be found around Dehradun, in Northern India, and at the Raccoon River in Des Moines, Iowa. In flat regions infiltration galleries have also been used for many decades, delivering the water to a collector chamber from where it is pumped to the waterworks.

The first waterworks in Dresden-Saloppe, built between 1871 and 1875 on the right bank of the Elbe River, has an infiltration gallery. Drain pipes were installed near the riverbank to abstract raw water. Due to geological boundary conditions, more than 90% of the abstracted water is bank filtrate (Fischer et al. 2006). In 2010 the waterworks is still in active operation producing up to 12,000 m^3/day for industrial water supply.

An infiltration gallery along the Parapeti River at the city of Camiri, Bolivia, has been in operation since the early 1980s, and no critical operation and maintenance problems have been encountered (Camacho 2003).

Along an 80 km strip of the River Danube, more than 700 wells are installed, many of them within siphon systems. A system at the Szentendre Island north of Budapest can deliver 730,000 m^3/day (Homonnay 2002); some wells have been operated for more than 80 years. Water is transported through a converted siphon beneath the river (Molnar 1941).

The Karany water works has been supplying the capital Prague, Czech Republic, since 1914. The well gallery is located parallel to the River Jizera at a distance of about 250 m. The wells are connected to a siphon pipe (PVK 2000).

When looking at the various types of purification processes intentionally allowed to occur in the aquifer during bank filtration and artificial infiltration, the question inevitably arises whether these processes will be exhausted or whether the underground passage can be used continuously. In Germany, bank filtration and artificial groundwater recharge have been used for the drinking water supply at several sites making use of different subsoil characteristics for decades and no loss of purification capacity could be noticed. Despite considerable quality fluctuations of surface waters, natural processes did always (even following temporarily massive interruptions) turn back to normal. Furthermore, no alarming accumulation of persistent pollutants could be ascertained by investigation of various subsoils used for underground passages (Schmidt et al. 2003).

5.12 Implementation, Challenges, Strategies

- What are the major challenges to the development and implementation of bank filtration systems?

One of the most difficult challenges related to planning and implementing bank filtration systems at new sites is the uncertainty, until and even after the facility is actually operating, regarding yield and water quality that will be realized from the RBF facility. This uncertainty translates to an assumption of risk by decision makers and project proponents when they are considering whether bank filtration is the most appropriate water-treatment strategy. Although it is not possible to eliminate this risk, it can be minimized by careful and thorough planning conducted in a phased manner focusing on the key issues described in this chapter. It is also helpful if there is precedent within the watershed or river system demonstrating that bank filtration can be successfully used locally as a means of water treatment. It is essential that project designers communicate, without overstating, the risk associated with implementing new RBF systems and manage expectations throughout the entire project.

- What are the regulatory challenges faced by RBF systems and what strategies are being taken to meet such challenges?

As described in Sect. 5.4, in the United States the determination of the regulatory classification of water treatment facilities utilizing bank filtration is a significant challenge and presents uncertainty for water utilities considering bank filtration. In particular the determination of whether or not an RBF facility is classified as a "groundwater" or GWUDI system is an important distinction possibly resulting in significantly higher costs for operation and the construction of additional treatment facilities.

Compounding this issue is the lack of a clear, scientifically based, consistent method of determining whether or not a system is GWUDI. Two reasons for the lack of a consistent method for GWUDI determination are (a) the mechanisms of natural filtration are not well understood, and (b) the natural complexity, heterogeneities, and site-specific conditions of alluvial systems make a "cookbook" or "one size fits all" approach impractical. Ongoing research to better understand the complex process of natural filtration should continue to enhance the scientific basis of regulatory policies. Regulatory programs, additionally, need to be developed in a more flexible manner to recognize and address the natural complexities and uniqueness of individual sites. Proper planning is critical to evaluate site-specific conditions.

- How can watershed-scale measures improve RBF operations?

The linkage of land-use planning with water-supply and water-quality criteria on a watershed basis can substantially increase the reliability of RBF facilities. Land-use practices reducing point and non-point pollutant discharges into surface water and groundwater will improve water quality and ease the strain on the natural alluvial filtering system. Some land-use practices—such as either the reduction or increase of discharged wastes from agricultural, industrial, mining (e.g., gravel), and wastewater operations have direct effects on water quality. Land-use practices reducing erosion and pollution from sediment will indirectly improve water quality. Examples of land-use practices having an indirect benefit on water supplies include those reducing erosion by regulating improved grading practices and those requiring timber-management practices designed to both reduce erosion from logging activities and minimize forest wildfires (increased erosion is one aftereffect of forest wildfires).

- What are the needs in developing or recently industrialized countries for the sustainable operation of RBF systems?

In developing countries such as China, Egypt, and India high levels of water pollution make it difficult for RBF systems to produce filtrate that can be supplied directly to consumers without the need of an additional treatment facility. The successful removal of water color and organic contaminants through pre-chlorination RBF is common in many locations. Rarely reported or recognized, however, is the positive effect RBF has on minimizing the formation of dangerous disinfection by-products. Following an initial chemical treatment of the water, many water utilities chlorinate treated water a second time before supplying it to the distribution system thus creating a significant potential for high levels of dangerous disinfection by-products.

Although RBF systems in developing or recently industrialized countries face major challenges from poor-quality source water at many sites (caused by poor upstream wastewater treatment practices), Ray (2008) described the worldwide potential for RBF. Sandhu et al. (2010) also described the challenges facing RBF systems in developing countries (e.g., India) with negligible controls over the discharge into rivers of raw industrial and municipal wastes.

RBF systems also face hazards from rivers with frequently shifting banks. Regulations for wellhead protection in upstream watershed areas are nearly non-existent in many developing countries. In such circumstances the entry of contaminants from groundwater into drinking-water supplies is often unavoidable.

RBF systems are rapidly gaining popularity in many recently industrialized countries. In one such country, the Republic of Korea, however, geologic deposits are frequently coarse, thin, and not suitable for installing vertical wells. Well-placement and pumping rates for horizontal collector wells also have to be carefully addressed so that river scouring does not cause turbidity breakthroughs. And, as elsewhere, industrial and municipal effluent discharges and low-flow conditions may cause significant organic loads generating anoxic conditions in rivers at current and potential Republic of Korea RBF sites.

- Is RBF adaptable to changing climate conditions?

Temperature increases generating higher global radiation and increased evaporation combining with reduced summer precipitation will lead to extended low-flow periods. Low-flow periods result in greater portions of wastewater in surface water, lower discharge rates, and thus a decreased contaminant dilution in source water. Increased concentrations of trace organic compounds such as cosmetics, ethylene diamine tetraacetic acid, pharmaceuticals, and radio-opaque substances that are not efficiently retained by common wastewater treatments are observed in surface waters subject to anthropogenic contamination. Higher global radiation, higher water temperatures, and abundant nutrient supply result in increased biomass production, which, in turn, causes a higher suspended-matter load. Together with the low flow velocities and reduced shear stress characteristic of low-flow periods, this leads to more intense clogging at the bottoms of water bodies (Schoenheinz and Grischek 2010). The infiltration rate and, consequently, the bank filtration portion in the raw water supply will be reduced. Increased microbial degradation of organic substances leading to source-water oxygen depletion, along with the lower solubility of oxygen caused by higher temperatures, may cause a shift from aerobic to anoxic conditions in the bank filtrate. This was seen in the Rhine River during the extreme-low-flow summer of 2003 (Eckert et al. 2008).

Floods caused by heavy rainfall are accompanied by the input of diffuse matter into various source waters from the flushing of temporarily flooded agricultural, industrial, and urban areas. Floods generate inputs such as increased bacteriological loads (eluted from grazing land or from sewer-system stormwater overflow), higher concentrations of dissolved organic compounds (of both anthropogenic and natural origin), insect repellents, nutrients from fertilizers, and pesticides. By producing higher flow velocities and higher shear stresses, floods also at least partially remove the clogging layer from river-bottom surfaces at many sites.

Boundary-condition changes at bank filtration sites as a consequence of possible climate change, together with their effects on RBF and possible adaptive measures, are discussed by Schoenheinz and Grischek (2010). The most effective tool for neutralizing fluctuations in organic-compound concentrations is maximizing travel time as a function of the flow-path length and the hydraulic gradient between the source and the abstraction well. Sufficiently long travel times to compensate for slower anoxic degradation, concentration variations, or desorption processes may be achieved by adequate selection of abstraction well locations (with the appropriate distance to the bank), site-specific optimization of abstraction quantities, and sufficiently long flow paths. Event-related flow-path observations by measurements taken at monitoring wells of the filtrate traveling between the river and the abstraction wells is critical for identifying potential breakthroughs of persistent organic compounds or pathogens. Planning for and installing emergency alternative water-treatment measures may be required.

One of the most significant advantages of RBF systems is that, relative to conventional water treatment systems, RBF systems are more capable of adjusting to changing conditions. Smaller snowmelts or increased seasonality of rainfall resulting from possible climate changes may require water-treatment facilities to adapt infrastructure and operations. RBF facilities are more amenable than conventional water treatment systems to conjunctive-use strategies that divert and treat water whenever it is available and then store that water in surface reservoirs or groundwater aquifers for subsequent use during periods of high demand. As both the initial capital costs for RBF systems and their operation and maintenance costs are low in comparison to conventional purification systems, RBF systems (which is considered a natural treatment system) offer real advantages for modifying operations to adapt to changing conditions.

Chapter 6
Solar Distillation

Rahul Dev and Gopal Nath Tiwari

Abstract "Solar distillation" is a technology for producing potable water from brackish and underground water of low-quality at low cost. It can reduce water-scarcity problems together with other water purification technologies. Solar distillation is analogous to natural hydrological cycle. It uses an apparatus called a solar still in which water is evaporated using solar energy, a form of renewable energy, and collected as distillate after condensation of the vapor. It effectively produces distilled water after removal of impurities. The major advantage of this is the use of solar energy instead of electrical energy generated from conventional fuels. This helps in producing potable water without degrading our environment. Over time, researchers have studied several designs of solar stills to evaluate its performance for different climatic, operational, and design parameters. In this chapter, many aspects of this technology are being covered. Among those, the working principle explains the production of distilled water through evaporation and condensation process inside the solar still, including its advantages and drawbacks. The evolution of the solar still is discussed under the historical background section and classified as passive or active solar stills. Various heat gains and losses and a methodology to develop a thermal model to predict the solar still performance are presented. Thereafter, the effects of various parameters, economic, energy-exergy analysis including carbon credit, and technology transfer with challenges in adoption are given. In conclusion, the probable extensive use of this technology to solve water problems on a large scale all over the world is presented.

Keywords Solar distillation • Potable water • Solar energy • Exergy

---

R. Dev (✉) • G. N. Tiwari
Centre for Energy Studies, Indian Institute of Technology Delhi,
Hauz Khas, New Delhi, 110016, India
e-mail: rahuldsurya@gmail.com, rahuldsurya@yahoo.com

G. N. Tiwari
e-mail: gntiwari@ces.iitd.ac.in

C. Ray, R. Jain (eds.), *Drinking Water Treatment,* Strategies for Sustainability,
DOI 10.1007/978-94-007-1104-4_6,

6.1 Introduction

Water is a precious resource for sustaining life. It is used in many areas such as irrigation, domestic and industrial fields, etc. But the availability of fresh water is fast decreasing all over the world due to mainly anthropogenic activities. Water scarcity is becoming a major problem for all nations. Most people do not have options other than using polluted water (water containing a large quantity of salts, solid matters, hazardous elements, compounds, etc., which is not suitable to use in any application without treatment) for living. Due to the rising problem of drinking water, developed and underdeveloped countries set norms for drinking water according to the salt content. Table 6.1 shows broadly classified norms of drinking water set by the World Health Organization (WHO) and different countries.

6.2 Water Characterization

Brackish water can be a better resource for getting pure water after treatment. This treated less saline water can be obtained through distillation. The amount of total dissolved solids (TDS) measured in part per million (ppm, also commonly referred to as mg/l) predicts the salt intensity of the water (Table 6.2).

- **Seawater:** Seawater contains about 50 simple substances, where chloride represents 55% from the total weight of the dissolved salts. However, seawater salinity is closer to 35 g/l, but it changes from one part of the sea to another part of the sea.
- **Brackish water:** Brackish water is defined as the water having salinity between 1 and 3 g/l (Sandec 1997). Although the salinity of this water is lower than that of seawater, it is not fit to drink and other uses.
- **Fresh water:** Fresh water contains TDS less than 1 g/l.

Table 6.1 Salt content for drinking water in different countries

Status of countries	Permissible limit of salt content (ppm)
Developed	500
Developing	1,500
Underdeveloped	2,000

Table 6.2 Water classification as per dissolved salt content (ppm conversion)

Water type	Total dissolved solids (ppm)
Fresh water	< 1,000
Brackish water	
Light	1,000–3,000
Moderate	3,000–10,000
High	10,000–30,000
Seawater	> 30,000

According to Abdenacer and Nafila (2007), the desalination process is possible for seawater from feed water with a salinity closer to 35 g/l, or for brackish water with a salinity between 1 and 10 g/l. However, the efficiency of desalination is different due to thermal properties of the source water, which depends upon the quantity of the salt content. Abdenacer also reported that brackish water (with less salt) requires less energy of any form for its desalination than that required by seawater.

6.3 Solar Distillation: Basic Principle

Solar distillation is a technology that uses solar radiation as a source of heat energy to generate distilled water. This technology uses the principles of greenhouse effect and evaporation/condensation of water. Solar distillation technology is a small-scale analogy of nature's hydrological cycle, which provides fresh water at a very large scale.

A solar still is a device used in which impure/saline water is fed to obtain distilled water by solar distillation. Figure 6.1 shows the schematic diagram of the conventional single slope passive solar still. It is a box type structure that can be made of materials such as fiber reinforced plastic (FRP), wood, concrete, or galvanized iron (GI) sheet covered with some insulation. The box is covered with a glass cover. The solar radiation passes through the glass cover. A major portion of this solar radiation is absorbed by the black painted surface of the basin, generally known as the basin liner. However, a small amount of reflection loss takes place at the glass cover, the water, and the basin liner surfaces. A small amount of solar radiation is also absorbed by the glass cover and water because of their absorptivity.

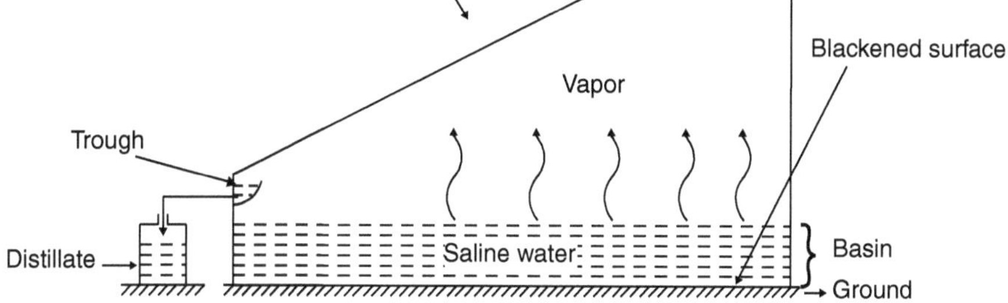

Fig. 6.1 Schematic diagram of a conventional single slope passive solar still

A large portion of heat absorbed by the basin liner is transferred to the saline water by convection effect as top heat loss. The remaining amount of absorbed heat is lost by conduction as bottom heat loss. Heat transfer from the water surface to the glass cover takes place by three mechanisms: evaporation, convection, and radiation. An evaporative heat transfer is a dominant effect, which is responsible for the production of the distillate. Evaporated water leaves most of the contaminants and microbes on the basin liner. Further, this evaporated water with some volatile material undergoes film type condensation at the inner surface of the glass cover. A film type condensation occurs because of inclination of the glass cover, cohesion between condensed water molecules, and gravity effect. The condensed water trickles down to a trough, which guides it into a container. The yield form a single slope passive solar still may vary from 0.5 to 1.2 l/m^2/day (in winter) and 1.0 to 2.5 l/m^2/day (in summer). And the overall thermal efficiency (the ratio of energy output to the energy input for a system, see Sects. 6.7.3 and 6.7.4) of the system can vary from 5 to 40% throughout the year. It is also seen that the salinity of the water is greatly reduced to 8–15 ppm.

There are various advantages of solar stills over other desalination technologies such as: (a) an easy, small-scale, and cost effective technique for providing safe water in homes or in small communities; (b) producing distilled water, (c) simplicity in design; (d) no moving parts (pumps, motors, etc.) are required to run the unit in passive mode of operation; (e) no conventional sources of energy are required which helps in reducing the environmental pollution as it requires only solar energy (low grade energy), which is renewable and non-polluting; (f) no skilled operator is required for operation and maintenance; (g) local manufacturing/repairing is possible; (h) purifying highly saline water (even seawater); (i) a sense of satisfaction in having their own trusted and easy to use water treatment plant on-site at home, for solar still users; and (j) effective removal of pathogens and chemicals reduces risk of heath problems associated with water borne diseases, etc.

However, a solar still has some drawbacks, which sometimes limit the use of this technique for large-scale production: (a) large solar radiation collection area requirements, (b) vulnerability to weather-related damage, (c) low yield, (d) low efficiency, (e) less market demand of technology, and (f) low interest of the manufacturers, etc.

The solar distillation technology has major application for producing potable water in marshy, coastal, rural, and remote areas, etc. (i.e., for on site production of the water where the electricity has not been reached due to which the implementation of other technologies is not possible and solar distillation can be the only option there for producing water). The distilled water produced by solar stills has many applications such as cleaning utensils, mechanical parts, and apparatus in chemical laboratories, educational institutes, automobile garage/workshops, hospitals, and industries. This distilled water can also be used in various chemical reactions. The distilled water is not fit for direct use in drinking and cooking applications because of the lack or absence of minerals that are highly desirable for the human body. However, we can prepare suitable water for these purposes after pre-treatment.

6.4 Historical Background: Evaluation Process of Solar Stills

It is believed that the greenhouse effect was first discovered by the Egyptians and that the use of solar energy began in the third century with Archimedes (in 214–212 BC) who made a heat ray in which he used solar energy to defend the harbor of Syracuse against the Roman fleet. Archimedes used a mirror to set fire to the ships of the Roman fleet while standing on the shore. In 1615, Salomon de Caux made the first solar engine using glass lenses, a supporting frame, and an airtight metal vessel containing water and air, which produced a small water fountain when the air heated up during operation. And in 1774, Joseph Priestley discovered oxygen gas through an experiment in which he focused sunlight on mercuric oxide (HgO) inside a glass tube.

The earliest documented work is that of an Arab alchemist in 1551 (Mouchot 1869). In his review, Mouchot writes:

> One uses glass vessels for the solar distillation operation… According to the Arab alchemists, polished Damascus concave mirrors should be used for solar distillation.

Nebbia and Menozzi (1966) mention the work of Della Porta (1589) in a review of water desalination, which was published in 1589. The solar still used by Della Porta is shown in Fig. 6.2. In his own description of the experiment, he said:

> Insert green leaves into wide earthen pots full of water, so that the vapors may thicken more quickly into water. Turn all this apparatus, when it has been very carefully prepared, to the

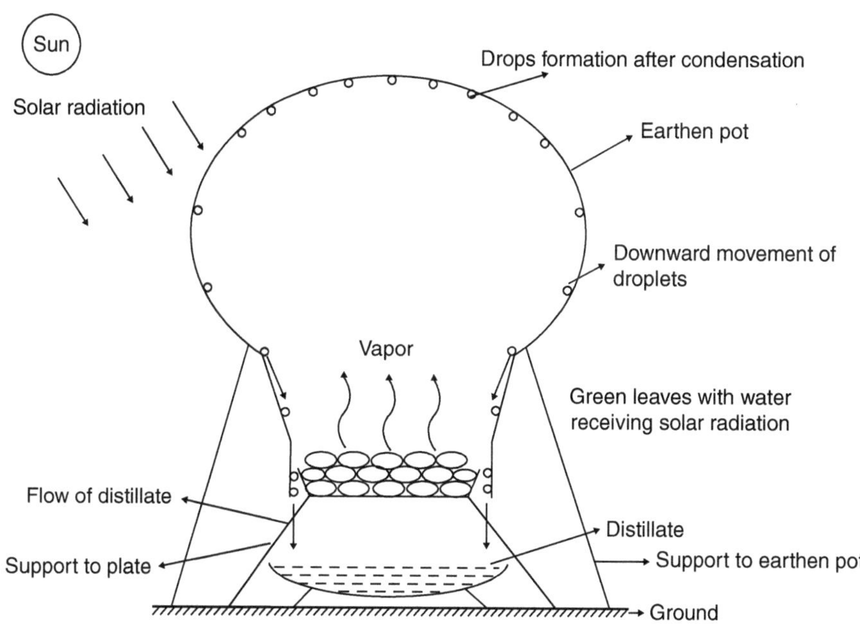

Fig. 6.2 Solar distillation apparatus of Della Porta

> most intense heat of the sun's rays, for immediately, they dissolve into vapors, and will fall drop by drop (i.e. the distillate which contains the essence of the green leaves etc.) into the vessel which have been placed underneath. In the evening, after sunset, remove them and fill with new herbs. Knot-grass, also commonly called "sparrow's tongue," when it has been cut up and distilled is very good for inflammation of the eyes and other afflictions. A liquid produced by ground-pine which will end all convulsions if the sick man washes his limbs with it; and there are other examples too numerous to mention.

The great French chemist Lavoisier (1862) used large glass lenses, mounted on supporting structures to concentrate solar energy on distillation flasks filled with content to be distilled. The use of silver or aluminum coated glass reflectors to concentrate solar energy for distillation purposes has also been described by Mouchot (1869). Thus, it is obvious that nineteenth century scientists took interest in using mirrors and lenses for harnessing solar energy for distillation.

The conventional solar distillation system, commonly known as the "solar still," was first designed and fabricated in 1872 near Las Salinas in Northern Chile by Carlos Wilson, a Swedish engineer. It was a large basin type solar still meant for supplying fresh water to a nitrate mining community. Several wooden bays of 1.14 m × 61.0 m were joined together with a total surface area of 4,700 m^2 covered with glass at top. The bottom of the bays, exposed to the sun, was blackened with logwood dye and alum. Brackish water was poured into the bays that, upon evaporation aided by solar energy, condensed over the glass cover and trickled down into the collectors. This device was in operation for about four decades (until 1910) and yielded more than 4.9 kg/m^2/day in summer (Harding 1883). It is worth noting that this output compares well with the distilled water output from the present day solar stills. The major problem with the still was the accumulation of salt in the basin, thereby, necessitating regular flushing of the still.

No work on solar distillation seems to have been published after the 1880s until the end of World War I because of slow progress in the solar energy research field. After this, with the renewal of interest, several types of devices have been reported, e.g., roof-type, V-covered, tilted wick, inclined tray, suspended envelope, tubular and air inflated stills, etc. The use of metal coated reflectors as solar concentrators for application in solar distillation has been described by Kausch (1920). Pasteur (1928) also used several concentrators to focus solar rays onto a copper boiler containing water. The steam generated from the boiler was piped to a conventional water-cooled condenser in which distilled water was accumulated. Abbot (1938) used cylindrical parabolic reflectors (aluminum coated surface) to focus solar energy on evacuated tubes containing water. He also devised a "clock-work" arrangement to track the motion of the sun. Although efficiencies as high as 80% could be achieved, the boiling of water in the tubes created some problems. During World War II, Telkes (1945) developed air inflated plastic stills for the US Navy and Air Force to be used in emergency life-rafts. The arrangement consisted of an inflatable transparent plastic bag inside which a porous felt pad was suspended with collector bottles placed at the bottom. Whenever it was to be used, the felt pad was saturated with seawater. Water evaporated from the pad because of incident insolation and condensed on the interior surface of the bag. Distilled water was collected in the bottles kept at the bottom. As many as 200,000 of these stills were used by the US Navy during the war.

The next stage was to improve the operating efficiencies of the various types of solar distillation devices. Forced air circulation was tried in stills to enhance the vapor condensation rate. Several investigators have attempted to make use of the latent heat of vaporization, in either multiple effect systems or for preheating the brine to increase the output of the stills. Several large-scale distillation plants and integrated schemes for combining electric power generation and desalination of water have also been suggested as a way of improving the overall operating efficiency of the plant.

The basin-type solar still, also known as the greenhouse roof, simple or conventional type, is in the most advanced stage of development. Several researchers have investigated the effect of climatic, operational, and design parameters on the performance of such a still. Cooper (1969, 1973) analyzed the performance of a basin type solar still and showed the maximum possible efficiency of the solar still can be 60%. Frick (1970) has also proposed a mathematical model for the still based on the thermic circuits and the Sankey diagrams; his analysis is based on the assumption of sine wave heat flow. Hirschmann and Roefler (1970) have considered periodic insolation in estimating the effect of heat capacity on the performance of the still. The periodic and transient analysis has also been presented by Baum et al. (1970), Nayak et al. (1980) and Sodha et al. (1980a, b), respectively.

Several other types of stills have also been recently proposed and studied. Oltra (1972) studied multi-effect stills. Howe (1961) and Akhtamov et al. (1978) proposed a tilted tray or inclined-stepped solar stills. Frick and Sommerfeld (1973), Sodha et al. (1981), Moustafa et al. (1979) studied the tilted, wick type and multiple wick type solar stills. Norov et al. (1975), Menguy et al. (1980) have studied the solar film covered still and wiping spherical stills. Selcuk (1971), Sodha et al. (1980c) studied the solar still greenhouse combination for the distillate and plant production at the same time. Soliman (1976), Malik and Tran (1973), and Malik et al. (1982) studied indirectly heated solar stills. Scientists have also studied many designs along with reflectors, effect of heat storing materials, climatic, operational and design parameters with modifications in the conventional type solar still for last 30 years with significant improvement in terms of yield, efficiency, and presented vast studies. During recent years, several active solar stills have been studied by Chaouchi et al. (2006), Nassar et al. (2007), Dev and Tiwari (2009, 2010), and Dev et al. (2011). Some other common but important and advance solar stills have been explained in the passive and active solar still sections.

6.5 Broad Classification of Solar Stills

Solar stills can be broadly classified according to mode of operation: (a) Passive solar stills and (b) Active solar stills. The following are descriptions of these still types.

6.5.1 Passive Solar Stills

A conventional single slope solar still (single effect) has already been explained in Sect. 6.3 (solar distillation: basic principle). Passive solar stills are modified according to a working principle to enhance their performance by making some changes in the basin design of single slope solar still. Modification can be done to enhance evaporation, condensation, increasing heat gain and reducing heat losses from the still. Some of the modified conventional solar stills are as follows.

6.5.1.1 Modified Single Slope Solar Still

Multiple-effect solar still consists of a number of transparent (except lower basin) basins, which contains the saline water. This effect is carried out by other basins placed one-by-one in a vertical column. The vapor from the first basin condenses on the lower surface of the second basin and transfers latent heat to the water in the second basin. Hence, transferring latent heat to the other basin placed above is called the process of multi-effect in solar stills. These are typical high performance stills and produce a larger quantity of distilled water for a given insolation in comparison to that of a single effect type of the same basin area. But the number of effects increases the cost of the design, as well as the addition of material cost. A multi-effect still (Cooper and Appleyard 1967) has been designed to have maximum utilization of heat energy and reducing the losses through multiple condensation and evaporation phenomenon in one single unit of the solar still. According to Lobo and Araujo (1977), the intermediate basins adds about 15–25% to the cost of the standard still per unit area. Hence, they developed a multi-basin, multi-effect solar still (Fig. 6.3) to overcome the problem of the cost in multi-effects. They also observed that the gain in the distilled water output of the double effect still over the standard still can be at least 40–55% at 20–25 MJ/m^2/day and the cost for a given yield can be 20–25% cheaper than that of multi-effect solar still. El-Sebaii (2005) worked on a triple-basin solar still and observed that on a typical summer day of Tanta (latitude 30° 47′ N), the total productivity of the still was 12.635 kg/m^2/day. Tanaka et al.

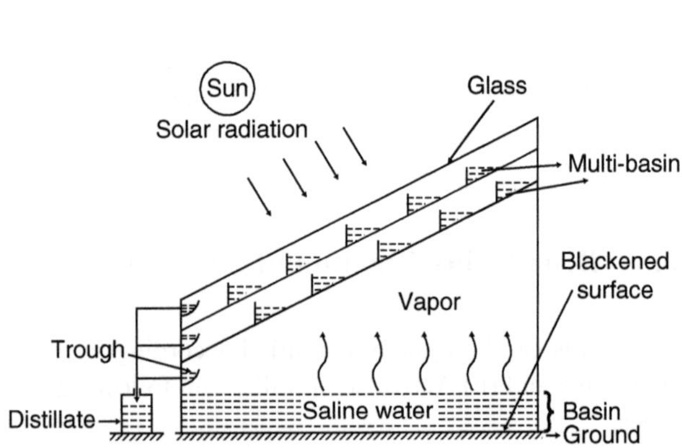

Fig. 6.3 Multiple-effect multi-basin still

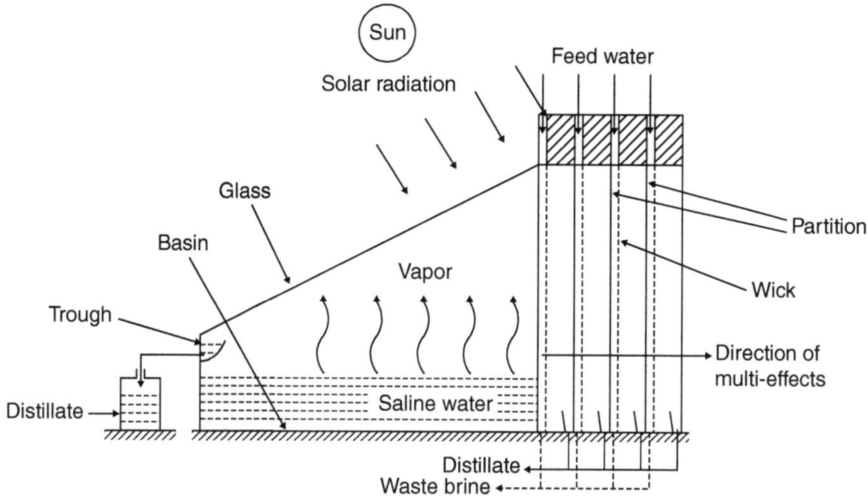

Fig. 6.4 Schematic view of a single slope, single basin, multiple effect solar still

(2000, 2002) improved the design of the single slope solar stills for higher productivity (Fig. 6.4). In this case, the outer portion of the vertical wall is covered with a wetted wick so that maximum condensation takes place on its inner surface due to its lower temperature in comparison to the temperature of the glass cover. The latent heat of condensation is released to the vertical walls. The vertical walls are heated along with the brine water trickling over it and further evaporation takes place, and then condensation on the other metallic wall with wetted wick after releasing the latent heat of condensation, and this process goes on.

A further modification of the single slope solar still is a regenerative type solar still. A regenerative solar still studied by Prakash and Kavathekar (1986) is very similar to the multiple-effect solar still in which the latent heat of condensation from the evaporated water is transferred to the water flowing over the condensing glass cover by conduction and convection. Flowing water over the inner glass cover is pre-heated before entering the still basin. Zurigat and Abu-arabi (2004) studied the modeling and performance of regenerative solar desalination systems and reported about 34–35% increase in productivity for the month of March and June for Oman weather conditions.

Another modification is an inverted absorber solar still. An inverted absorber solar still (Fig. 6.5) is a concentrating type of single slope solar still. It consists of a conventional solar still and a curved reflector below its metallic basin plate. An inverted absorber solar still receives solar radiation from the top as well as bottom unlike the conventional solar still. The addition of a reflector below the basin concentrates the solar radiation on the lower surface of the metallic basin. Because of this additional heat, the temperature of the water in an IASS increases at a faster rate in comparison to the conventional solar still. Tiwari and Suneja (1998) also reported the theoretical yields 5.1, 4.8, and 4.5 l/m^2/day at different water depths 0.01, 0.02,

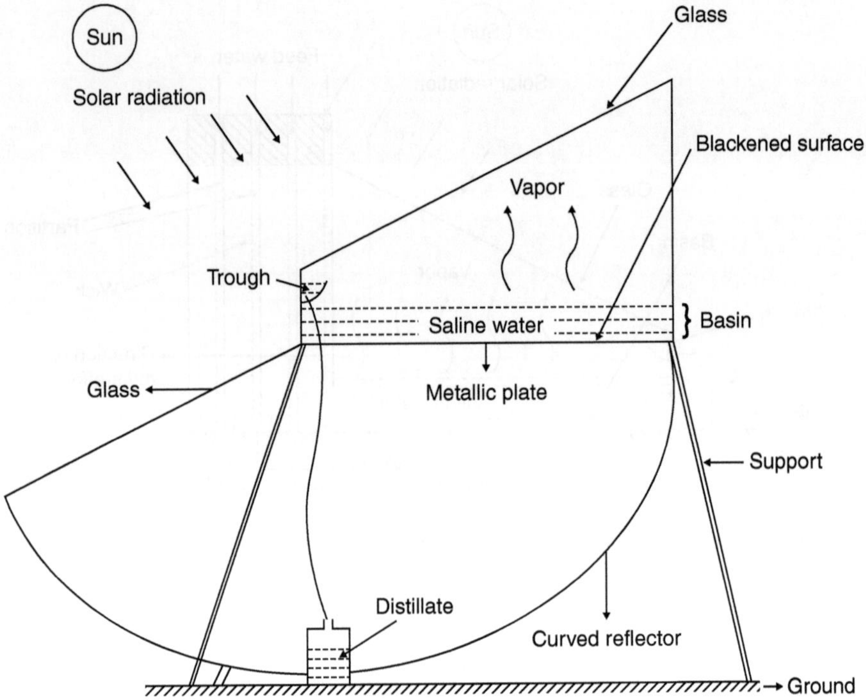

Fig. 6.5 Cross-sectional view of inverted absorber single slope solar still

and 0.03 m, respectively, for the single basin IASS for the composite climate of New Delhi, India.

In another modification, a baffle absorber plate was placed to float over the water surface inside the solar still (El-Sebaii et al. 2000), as shown in Fig. 6.6. An aluminum sheet painted black at the top surface and thermally insulated at the bottom—called a baffle—is used to float below the water level. An aluminum plate divides the unit into a shallow basin evaporating zone above the plate and a heat storing zone below the plate. During the sunshine hours the plate absorbs part of the solar energy that enters into the whole system. This absorbed heat increases its temperature and is conducted to the relatively lower temperature brackish water in the heat storing zone, below the aluminum plate. The hot water remains above the plate due to its lower density, therefore no heat transfer takes place by convection between the two zones through the 15 mm gap left between the plate and the unit sidewalls. Because the aluminum plate is insulated at the bottom, the heat transfer between the water above and below the plate by conduction is very small. During the night, the temperature of the water above the aluminum plate falls below that of the heat storing zone. Due to this temperature difference, convection currents gradually and continuously replace the relatively cold water above the plate by that of the hot water below the plate through the 15 mm gap, and the evaporation and condensation processes continue until the temperature difference between the water and glass

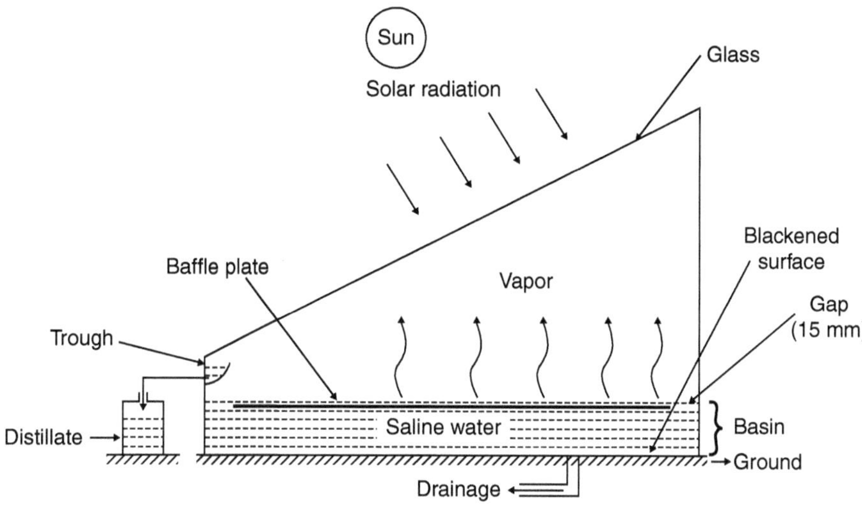

Fig. 6.6 Cross-sectional view of single basin solar still with baffle absorber plate

is maintained during nocturnal production. The next day, all the processes are restarted to produce distillate. El-Sebaii et al. (2000) found that the daily productivity of this modified solar still is about 20% higher than that of the conventional single slope still for the climatic condition of Tanta, Egypt.

6.5.1.2 Double Slope Solar Still (DSSS)

A double slope solar still with different sun positions over the still is shown in Fig. 6.7. It has two condensing glass covers facing east and west and a basin filled with saline water. A single slope solar still has the drawback of harnessing solar radiation for a limited time during the day, i.e., when the sun position is such that

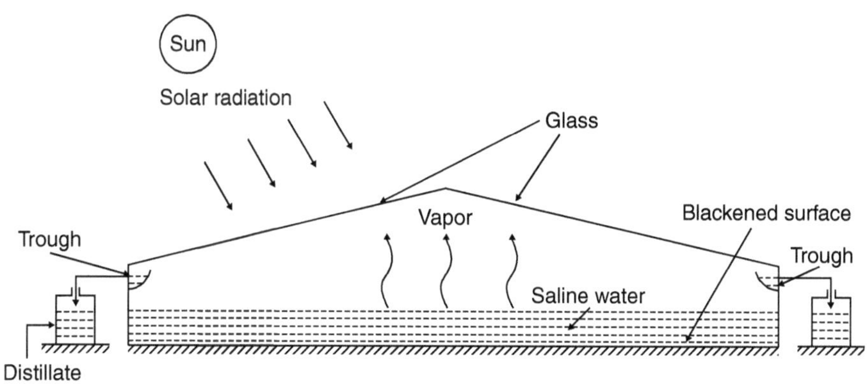

Fig. 6.7 Double slope single basin solar still

the solar radiation incident is normal or close to normal to the glass cover. A double slope solar still also works on the same principle as a single slope solar still but with the difference of heat input over the entire day. A double slope solar still is comparatively more effective in terms of yield/cost and hours of operation though annual yield is slightly lower than that of single slope passive solar still. Yadav and Tiwari (1987) made a comparative study of the performance of single and double slope solar stills. It was concluded that the yield could be higher in winter for single slope stills and higher in summer for double slope stills. Rubio et al. (2004) have also studied the performance of double slope solar stills for the climatic conditions of Mexico and developed a thermal model, which has been validated by experimental results. In a modification to the double slope solar still, the provision for rain water collection can be made by extending the height of the solar still walls in design to make it economical also. During rainy days, the absence of solar radiation is likely to happen due to clouds, the solar still becomes non-functional. At that time the rain water can be collected because the rain water is already distilled water.

6.5.1.3 Multi-wick Solar Still

A basin type solar still has a drawback in the storage of heat due to the heat capacity of water. Because of this, it takes time to increase the water temperature by storing the heat and then for evaporation of water to take place. It also leads to nocturnal (during the night time) production of distilled water. To overcome this drawback, Sodha et al. (1981) proposed a multi-wick solar still. In this design the heat storage effect is minimized because of the continuous flow of saline feed water, i.e., by making a water film through the capillary effect (Sodha et al. 1981; Tiwari and Salem 1984). Water, fed at a slow rate through the porous absorbers, gets evaporated immediately after receiving the solar radiation incident of the absorber and condenses on the inner surface of the glass cover. A water reservoir is introduced at the top of the absorber and the ends of a number of jute cloth pieces (porous fibers works as a high capacity capillary) are dipped into the reservoir while the other ends are spread over the absorber (Fig. 6.8). Each jute cloth is separated by a thin sheet of polythene acting independently and sucking water (capillary action) from the reservoir to the inclined surface tilted at such an angle as to intercept maximum solar radiation. The number of jute cloths used for this depends on the porosity of the fiber and the water level in the reservoir. In a later development, a single slope multi-wick solar still was converted to a double slope multi-wick solar still (Tiwari and Tiwari 2007c). Singh and Tiwari (1992) studied double effect multi-wick solar stills to increase the still efficiency by utilizing the latent heat released by the vapor at first effect. Kumar and Anand (1992) studied the tubular multi-wick solar still and showed that it gives distillate output of about 8, 13, and 18% more than the tubular solar still, the simple multi-wick solar still and the conventional basin type solar still, respectively.

It has been observed that the average daily distillate per m^2 from double slope still (east-west) is higher than that of a single slope facing south in the summer as

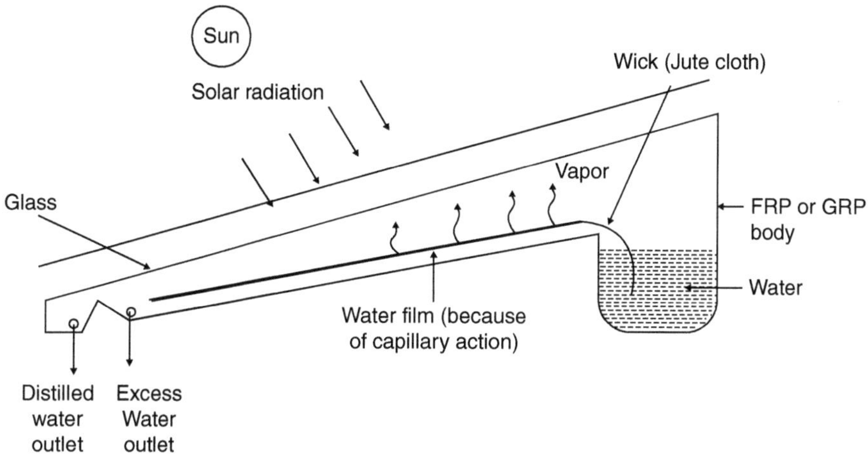

Fig. 6.8 Cross-sectional view of FRP multi-wick solar still

the total incident of solar insolation in the former still is higher than the latter due to overhead movement of the sun (looking from the Earth). This design has the following advantages over the conventional solar still: (a) it is possible to choose and set the still in any direction at any angle to receive maximum solar radiation because of the less bulky design than basin type solar stills, (b) there is no shadowing effect due to the small height of the side walls, and (c) the sediment deposits on the blackened cloth can be brushed off easily when necessary.

There are a few limitations also associated with the multi-wick still: (a) the wick becomes dry during peak sunny hours, i.e., the drying rate is much faster than that of capillary action rate, (b) heat loss occurs through the excess of hot water that goes as waste to the drainage during late and early working hours because of the slow rate of evaporation in those hours, and (c) salt scale formation in the basin. These problems can be solved by the higher level of water, which always has the wick floating over the water reservoir; and feeding the excess hot water back into the reservoir during early and late working hours, i.e., recirculation of the water into the reservoir to utilize the heat in the hot water (Janarthanan et al. 2005, 2006).

6.5.1.4 Vertical Solar Still

A vertical solar still consists of a vertical glass frame with a hanging black wick inside. The brackish water is allowed to flow into the vertical solar still through a flow regulating valve onto the wicks. The water flows over the spongy absorbing surface (heated by the solar radiation) through a pipe distributor. A part of this water evaporates and becomes a distillate after condensation over the vertical glass and the rest of the water drains out through tubes with some elevated temperature. The distilled water is collected in a jar. The vertical partitions have minimal deformation and any

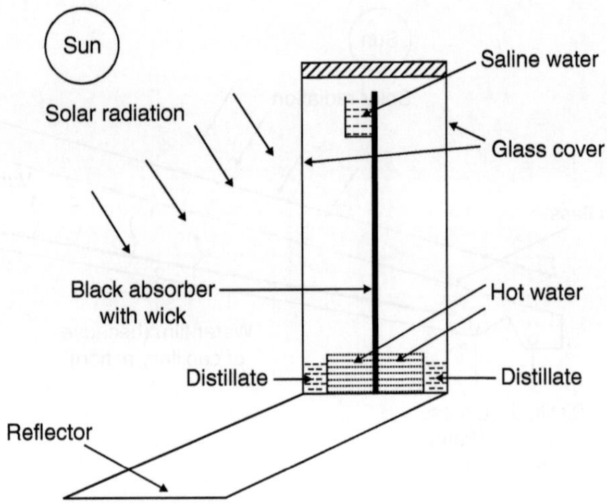

Fig. 6.9 Schematic diagram of a single effect vertical solar still

fibers protruding from the wicks absorb saline water, which becomes heavier and are gradually suppressed by gravity. The vertical solar still is most suitable in urban/rural areas where the land cost is very high since it occupies less horizontal area. The cost of distilled water is low compared to the conventional solar still. Figure 6.9 shows a vertical solar still with a mirror reflector.

For clear summer days in Algeria, Boukar and Harmin (2005) observed that the yield can be about 0.2725–1.31 l/m^2/day for a corresponding 8.42–14.71 MJ with an overall thermal efficiency from 7.85 to 21.19% which is lower than the conventional single slope solar still, i.e., 15–40%. Tanaka and Nakatake (2005a, b; 2007) studied a vertical multiple-effect diffusion-type solar still coupled with a flat mirror. The latent heat of vaporization available from the first effect is used to heat the water in the second partition and so on. It was seen that the daily yield increases linearly with up to four effects and then becomes non-linear. The daily productivity of the vertical multiple diffusion solar still with flat plate mirror was observed more than 30 l/m^2/day throughout the year at any latitude except for winter season at 40° N latitude.

6.5.1.5 Inclined Solar Water Distillation System

An inclined solar distiller based on capillary effect is shown in Fig. 6.10. This type of distiller (Bouchekima et al. 2001) uses a very thin fabric comprising a single finely woven layer, instead of a thick spongy fabric. This fabric is held in contact with the overhanging plate through the interfacial tension, which is much greater than the force due to gravity. Because of this, the fabric does not sag inside and remains stretched. Due to the Marangoni effect (Ramadane et al. 1986), the stabilization of a thin flowing film undergoing evaporation tends to tear and reduces to a set of independent rivulets separated by dry zones. In the inclined stills, contamination often occurs because of the saline-soaked wicks placed above the condensing surface and

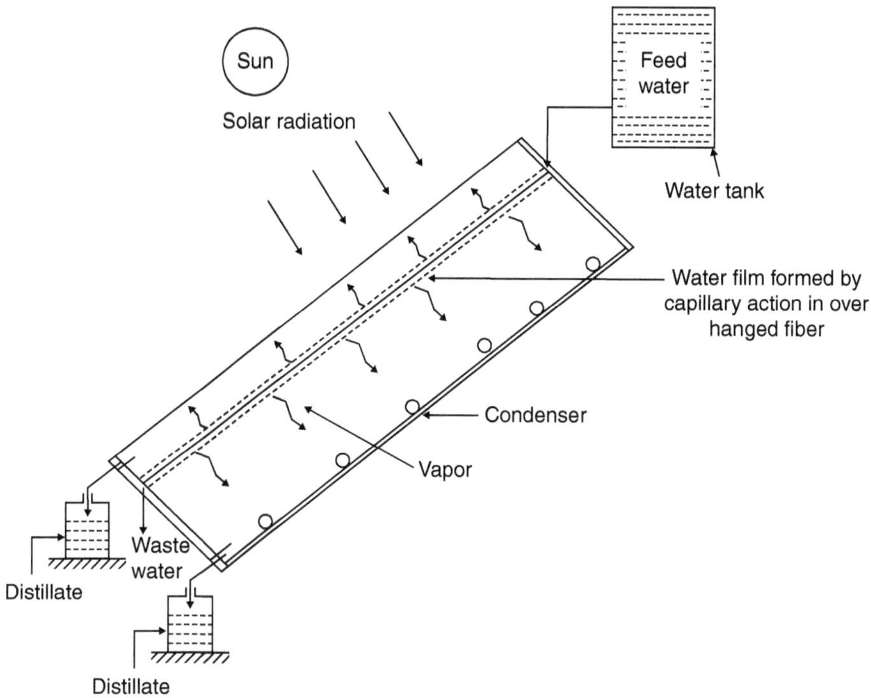

Fig. 6.10 Schematic diagram of capillary film distiller

the contamination may be caused by deformation of partitions due to gravity and fibers protruding from the wicks that are touching the facing and condensing surface. However, formation of the water film continues because of water flow, evaporation, and increased feeding of water. It is observed that maximum distillate 1.3 l/h/m^2 is obtained between 11 am and 12 noon at Sidi Mahdi in South Algeria.

The inverted trickle solar still (similar to Fig. 6.10), proposed by Badran and Hamdan (1995) and Badran et al. (2004), which was experimented for climatic condition of Amman, Jordan. The saline water is allowed to flow on the backside of an absorber plate. The water is kept on the plate with the help of a porous material fixed on its back. The rest of the body is made of galvanized black steel. Saline water enters via the inlet header such that it would uniformly spread beneath the plate's lower surface. A wire screen-jute sandwich is fixed to that surface to keep saline water on it. A low water flow rate is maintained such that its temperature should rise enough to produce vapor. The heat exchanger is made of copper tube. Condensation occurs in another compartment where distillate is obtained with additional product, i.e., water with reduced salinity, at the same time from the intermediate header and its production is found at 4.89 l/day. The productivity of the still is improved by adding a heat exchanger in the condenser up to 5.9 l/day. By reducing the salinity from 35,000 to 6,000 mg/l, the productivity of condensate increased from 2.5 l/day to 2.8 l/day. A large amount of the production (8 l/day) of water of reduced salinity

Table 6.3 Concentration and flow rates of various types of water flows in and out from the inverted trickle absorber solar still at 0.7 g/s

Water type	Amount of distillate (about) (l/day)	Concentration (mg/l)
Saline inlet	17	6,000
Saline outlet	10	6,272
Intermediate	5	3,600
Condensate or distillate	3	257

(2,424 mg/l) is possible by mixing the yield of the condensate and the intermediate headers (Table 6.3).

Aybar (2006) studied inclined wick type solar still (Fig. 6.11) and concluded that the daily yield of such solar stills is 3.5–5.4 l/m^2/day for summer climatic conditions of Turkey. The average water temperature available from such a solar still is around 40°C, which can be used for domestic application in addition to distilled water. Sadineni et al. (2008) studied an inclined weir type solar still for the month of September in Las Vegas and found 5.5 l/m^2/day with 20% improvement in yield over a single slope solar still.

6.5.1.6 Conical Solar Still

In a conical solar still (Tleimat and Howe 1967) the evaporating area is made bigger for a faster and higher condensation rate because the maximum heat is lost through

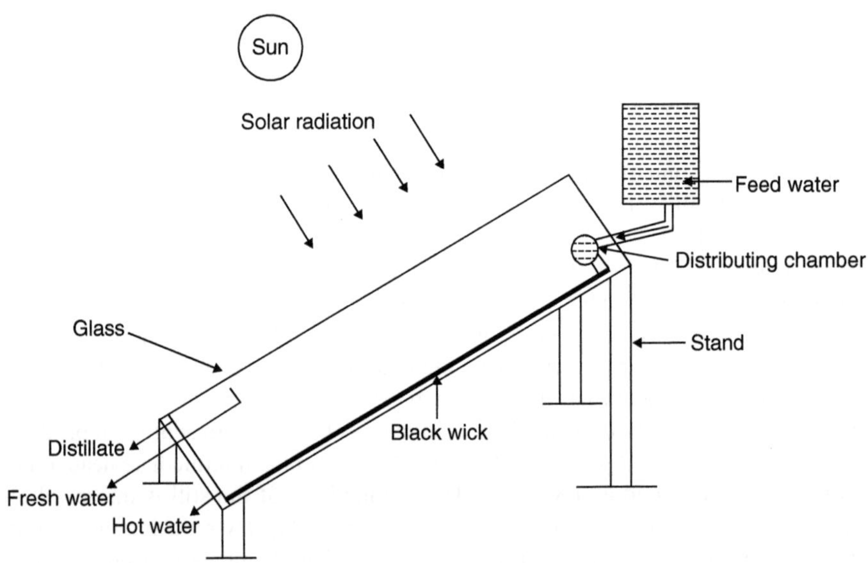

Fig. 6.11 Schematic diagram of inclined solar water distillation system

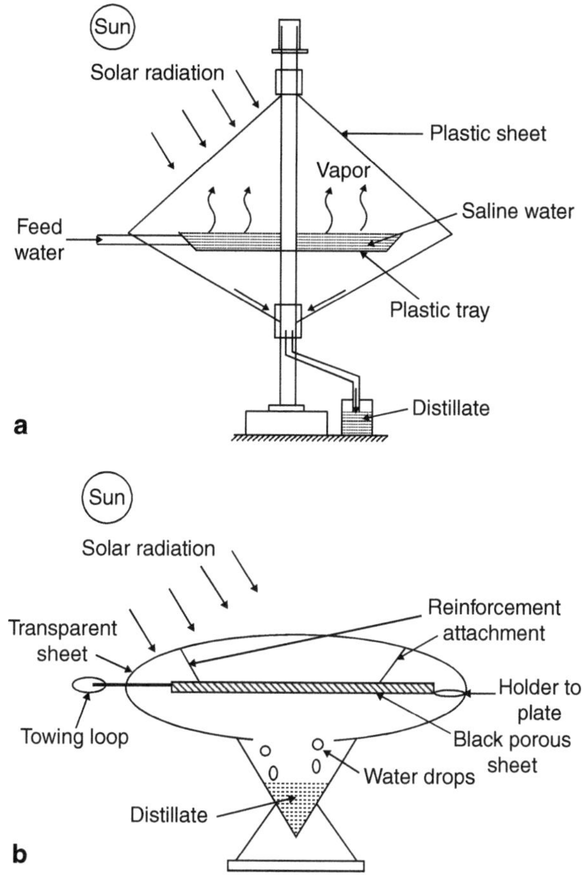

Fig. 6.12 Schematic diagram of **a** a conical solar still, and **b** life raft-type solar distiller

the condensing surface, i.e., plastic sheet (Fig. 6.12a). Impure water is enclosed in a transparent twin-cone arrangement. Solar energy trapped within the enclosure heats the water as usual, which causes evaporation and then condensation on the inner face of the transparent upper cone. The droplets of condensed water slide past into the water pan and are collected in the bottom cone. A conical solar still can be used for temporary installations of 4–5 months without maintenance costs. For a longer operation, it would be necessary to replace the cone sheet periodically. Because of its twin-cone configuration and the albedo effect (reflection of solar radiation from the ground) in snowy regions, this solar still is more efficient in polar areas. Other advantages include automatic sun tracking, the absence of orientation problems, and non-requirement of thermal insulation. A conical solar still yields 40% more distilled water than that of a conventional single slope solar still for the same surface area. Enhancement of the performance in tropical areas using the albedo effect can be achieved by making the floor surface relatively more reflective. The use of a dish with a parabolic surface may also improve the performance (Tiwari and Tiwari 2007b).

6.5.1.7 Life Raft Type Solar Still

An inflatable, floating type of solar still made of plastic for use with a life raft in the ocean was designed for the US Navy during the Second World War by Telkes (1945). The device (Fig. 6.12b) has no metallic or rigid parts and can be folded into a small volume of about one liter and weighs about 0.5 kg. Because of the bulkier designs of other solar stills, this solar still is very handy and transportable because it is made of a transparent plastic envelope in which a black porous pad is suspended to soak the seawater. This is a very useful solar still to use on ships at sea. The seawater, not fit to drink, can be used as feed water. The porous pad is supplied with the seawater and the whole assembly is exposed to solar radiation. The solar energy absorbed by the porous pad leads to the evaporation of water. The vapors condense on the inside surface of the plastic envelope and then trickle down into a collector bottle at the base of the still. Thus, one only has to inflate the plastic solar still, soak the pad in seawater, and set the still afloat beside the raft to get the potable water from it. A ballast tank filled with seawater is normally used to stabilize the floating still. The porous pad has to be flushed off periodically in seawater to wash out the excess salt deposited in it. In this case, the condensation is faster due to the large condensing surface area.

6.5.1.8 Film Covered Solar Still

Norov et al. (1975) suggested that an inflatable hemispherical plastic envelope to make a very sturdy, strong, and transportable design instead of the basin type solar still as shown in Fig. 6.13a. This solar still works on the same principle as the basin-type still. The condensed water trickles down to the lower sides of the envelope into a collector along the rim of the still. Another variety of polyethylene sheet (treated) (trade name "Sun Clear") was suggested by Umarov et al. (1976). A serious problem with this type of still was, of course, the deterioration of the plastic cover. However, the treated films did not deteriorate as rapidly as the untreated films. It was observed that stills covered with the sheets yielded about 4.2 l/m^2/day of distilled water in July at Tashkent, Uzbekistan, whereas the still with untreated film yielded only 2.7 l/m^2/day.

6.5.1.9 Wiping Spherical Still

This spherical type of still is similar to the raft-type still of Telkes (1945), with the difference that a blackened metal basin replaces the porous pad and an electrically operated wiper is attached to the top cover. The droplets condensed on the interior surface of the upper hemisphere are wiped by the electrically operated wiper. This way, the condensate is directed into the lower hemisphere through the gap between the basin and the enclosing sphere made of glass. The system developed by Menguy et al. (1980) is illustrated in Fig. 6.13b. The production of a 25% increase in distilled

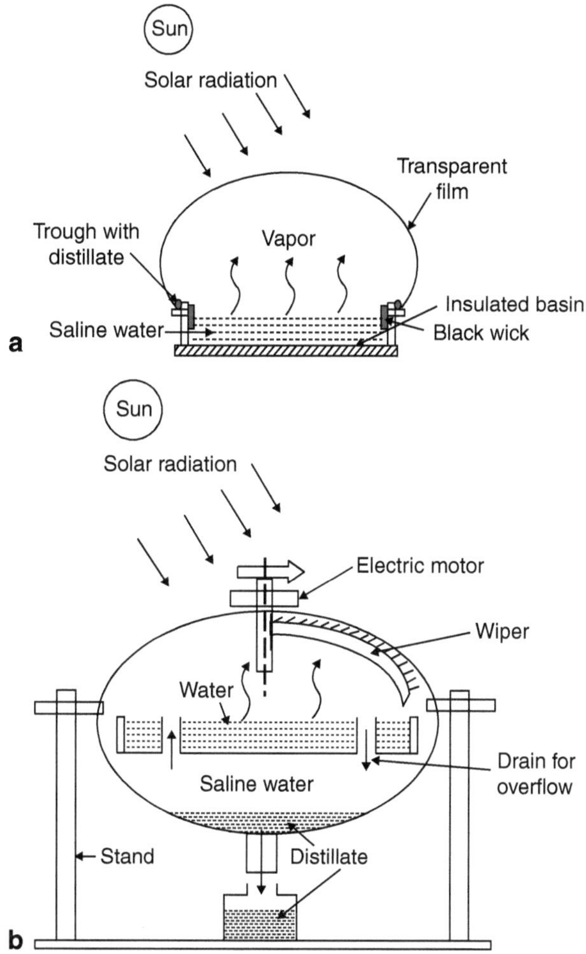

Fig. 6.13 Cross-section of **a** film covered solar still, and **b** wiping spherical still

water can be achieved by wiping the inner surface of the upper hemisphere because the cover is maintained uniformly transparent to allow maximum solar radiation to pass through and also dripping of distilled water back into the basin is prevented.

6.5.1.10 Concentric Tube Solar Still

Falvey and Todd (1980) have fabricated a new type of still using two concentric tubes (Fig. 6.14). The larger diameter tube is made of transparent flexible plastic. The smaller diameter tube is made of metal and blackened to act as an absorbing surface. This inner tube is open at one end. Saline water flows through a pipe wound along the entire length on the outside surface of the inner tube; the inlet and outlet for the saline water is at the same end of the still. Water oozes out of the pipe from

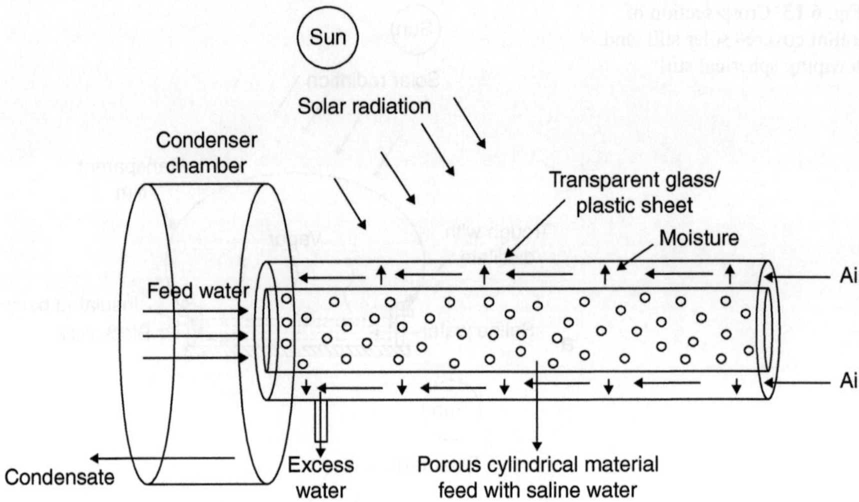

Fig. 6.14 Schematic diagram of concentric tube solar still

perforations made on it. Air at ambient temperature and humidity is blown through the annular space between the two tubes. The air becomes hotter because of the solar radiation absorbed by the inner tube. Consequently, it picks up moisture as the saline water oozes out of the piping. The warm humid air then flows into the interior of the inner tube and condenses on its cool inner surface. The air is then exhausted into the atmosphere and the distilled water is collected. The latent heat given at the walls of the inner tube preheats the saline water flowing in the perforated piping.

6.5.1.11 Solar Earth-Water Still

During the rainy season, a large quantity of moisture accumulates in the ground and this is recycled by the natural hydrological cycle with the return of this moisture to the atmosphere during the hot and dry climate. The earth-water still is made to capture this moisture to produce distilled water (Fig. 6.15). It is similar to the basin-type solar still with the ground replacing the basin. A pit is made in the ground covered with transparent material (e.g., glass or plastic sheet). Solar energy, transmitted through the transparent cover, heats the soil beneath the cover and vaporization of the moisture in the soil takes place. The vapors condense on the inner surface of the cover and the distilled water is collected in the collector-channel. This type of earth-water still has been studied by Kobayashi (1963) and Ahmadzadeh (1978). It was observed that no distilled water yield occurred when the system was just placed over the ground. However, if the still (area 0.5 m^2) was placed about 8 and 28 cm in the ground, where the corresponding moisture contents were of the order of 9 and 11.2%, respectively, 0.37 and 0.5 kg of distilled water could be collected within 24 h. The daily output from any such still reduces as the number of days increase

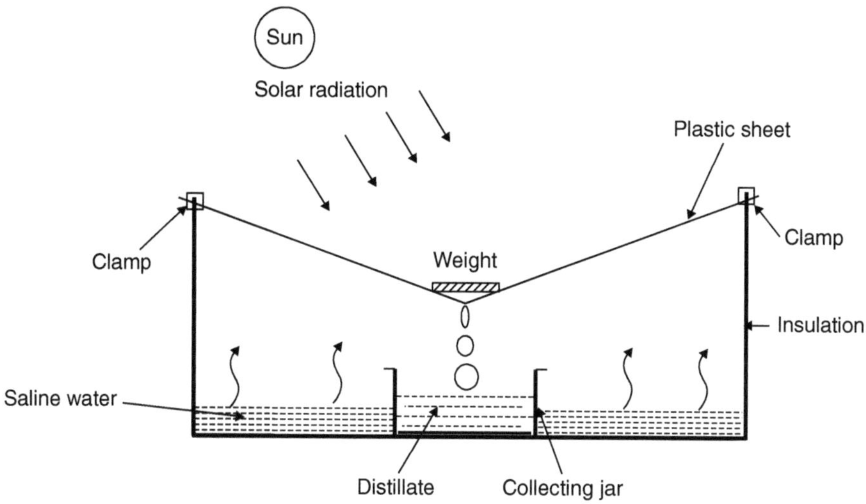

Fig. 6.15 Solar earth-water still within the form of a pit in the ground

because the moisture content of the soil decreases with the passage of time. This type of still may be an attractive emergency source of fresh water for drinking in the arid regions.

Murase et al. (2006) developed earth tube-type network solar stills for deserts for directly irrigating the roots of plants (Fig. 6.16). The upper part of the transparent polyvinyl chloride (PVC) outer tube-type solar still is used for the transmission of solar radiation to heat the brine (saline water) placed over a half-cut gray PVC tube. The lower part of the tube type solar still is buried inside the ground with a hollow tube to transfer distilled water to the root of the plants. After the transmission of so-

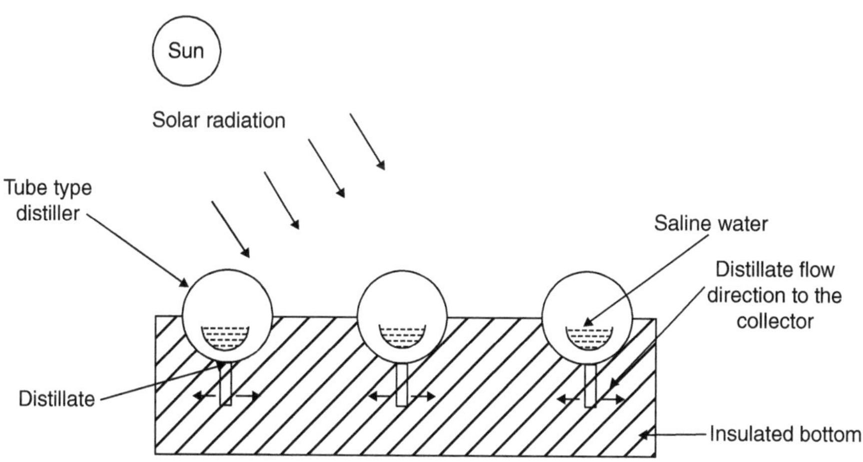

Fig. 6.16 Cross-sectional view of a tube-type solar still

lar radiation from the top of the solar still, it is absorbed in the half-cut tube and the water is heated. The heated water evaporates and then condenses inside the cover of the half-cut outer portion of the solar still. The condensed water trickles to the lower portion for irrigation. As expected, indoor experiments conducted earlier showed that at different water depths and irradiance, the yield and efficiency is higher at lower water depths. Further it is to be noted that there is linear dependency of yield with irradiance for lower water depth. The maximum efficiency of such a solar still is about 15%.

6.5.2 Active Solar Still

The daily distillate output of passive solar stills can be improved by increasing the temperature difference between the evaporating and the condensing surfaces. The water temperature in these stills mainly depend on the level of incident insolation along with water depth. The temperature of the evaporating surface can be raised by feeding the thermal energy into the basin by using some external sources which can be used for: (a) nocturnal production (utilization of stored energy during off-sunshine hours for evaporation by solar stills to produce yield is called *nocturnal distillation*): the feeding of hot water into the basin is done once a day, (b) pre-heated water application: the feeding of hot water into the basin at constant flow rate, and (c) high temperature distillation: the feeding of hot water into the basin from a solar collector panel.

Most of the solar stills discussed in the passive solar still section, such as single/double slope, vertical, inclined, etc., can be integrated with a flat-plate collector (FPC), an evacuated tubular collector, a parabolic concentrator, a heat pipe, a heat exchanger, and other devices to raise the temperature and flow rate of the feed water. This type of integration is called active solar distillation.

The active distillation is best suitable for commercial purpose because of the high initial investment, and high operating and maintenance costs. A few applications of active distillation are: (a) extraction of the essence from medicinal plant, and (b) unused hot water used for domestic purpose.

6.5.2.1 Flat Plate and Evacuated Tubular Collector Integrated Solar Stills

Soliman (1976) proposed integration of the flat-plate collector with solar stills for high temperature distillation. In this design, a single slope passive solar distillation unit is coupled with a flat-plate collector to heat up the water additionally (i.e., in the range of 70–80°C) prior to sending it into the solar still in the natural mode of operation. The water in the basin is heated: (a) by the solar energy directly received through the glass cover of the still, and (b) the water receives solar energy when it passes through the piping of the flat-plate collector beneath the absorber plate. Badran et al. (2005) also studied the performance of a single slope solar still connected with a single flat-plate collector and observed that there is significant in-

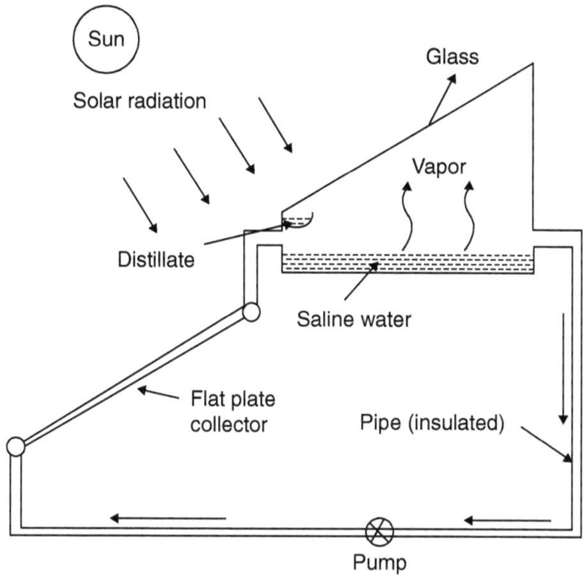

Fig. 6.17 An active solar still coupled with a flat-plate collector. (Kumar and Tiwari 1996a)

crease in daily yield, i.e., 36% compared with a passive solar still alone. Later, Rai and Tiwari (1982) developed a thermal model of the system and validated it through experimentation (Fig. 6.17). The performance was further increased by water flow over the glass cover, i.e., through regeneration (Bapeshwar and Tiwari 1984; Dutt et al. 1988). Kumar and Tiwari (1996a, b) optimized the number of collectors for maximum distillate output, which depends upon the number of effects. It has been reported that for a single effect, the optimum number of collectors can be 10, for double effect it may be 9 and for triple effect it can be 6.

In a hybrid photovoltaic thermal (PV-T) active solar still (Kumar 2008), a PV module (glass-to-glass) is used to generate electrical power to run the water in forced circulation mode by using a DC (60 W) water pump and to utilize the thermal energy to heat the water through the non-packing area of the PV modules. The effective area of the PV module was taken 0.605 m^2 with 36 solar cells that were connected in series. The PV module (75 W) was integrated with the flat-plate collector (2 m^2) by replacing the lower portion of the glass by the same. The advantage of integrating the PV module (glass-to-glass) with an active solar distillation system is to make this technology self sustainable to work in any remote area where there is no electricity, for commercial purposes. Kumar et al. (2008) reported the yield obtained by a hybrid PV-T active solar still could be 5–6 $kg/m^2/day$ for the summer conditions of New Delhi.

Tanaka et al. (2004, 2005) have also studied the performance of vertical multiple-effect diffusion-type solar stills coupled with a heat-pipe collector. The evaporation tubes of the solar collector are attached under the surface of the collector plate. The condensing tube is attached to the front surface of the first partition. This forms a closed loop between solar collector and condensing surface. The latent heat of vaporization of the first plate is used to heat the water of the second partition and so on.

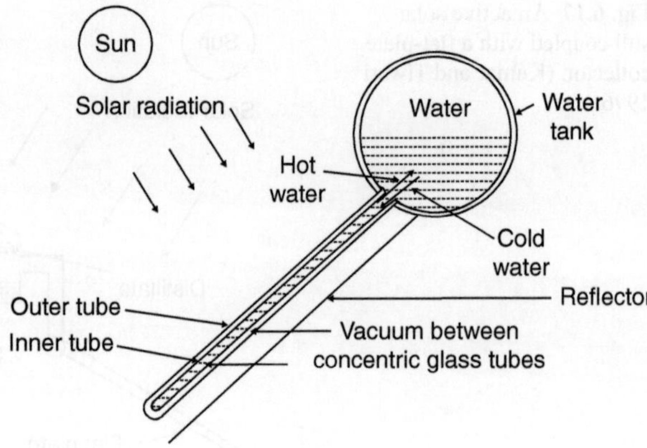

Fig. 6.18 A schematic diagram of evacuated tubular collector which can be integrated with solar still

Further, the FPCs are replaced by an evacuated tubular collector (Fig. 6.18) in later development of active solar distillation. Unlike FPCs, an evacuated tube collector has two concentric tubes with vacuum in the annular space between them (Budihardjo et al. 2007). This prevents mainly convection heat loss that occurs inside the collector due to air movements between the top glass cover and the absorber of a flat-plate collector to improve performance significantly. Although evacuated tube collectors are considerably more expensive than typical flat-plate collectors, they are much more efficient when high collection temperatures are needed for operating absorption. These collectors are more efficient than flat-plate collectors because: (a) they perform well in both direct and diffuse solar radiation; (b) the vacuum minimizes heat losses to the ambient, making these collectors particularly useful in areas with cold and cloudy winters; (c) the circular shape of the evacuated tube causes the sunlight to be received perpendicular to the absorber for most of the day and hence no tracking is required; whereas, in the case of a flat-plate collector in a fixed position, the sun is only perpendicular to the collector at noon; and (d) the evacuated-tube collectors achieve higher temperatures and efficiencies than flat-plate collectors.

6.5.2.2 Distillation Unit with Concentrator

Chaouchi et al. (2006) described a distillation unit integrated with a parabolic solar concentrator made of molded fiber glass (Fig. 6.19). The concentrating type solar distillation requires a continuous tracking of sun for solar radiation throughout the day. The sun tracking mechanism for this solar distiller has two axes according to Nuwayhid et al. (2001) and Kaushika and Reddy (2000). This dish surface is covered with stainless steel sheet segments of rectangular shape. At its focus is mounted the absorber, which is shaped like a cylindrical vessel. This absorber is completely insulated except the part that is lit by the solar rays reflected by the parabolic surface. The brackish water supply to the absorber is made in a continuous way in order to keep a constant volume of water in this absorber. The steam produced passes in a coil condenser where it will condense. The cooling water circulates with

Fig. 6.19 A schematic diagram of parabolic solar concentrator for distillation

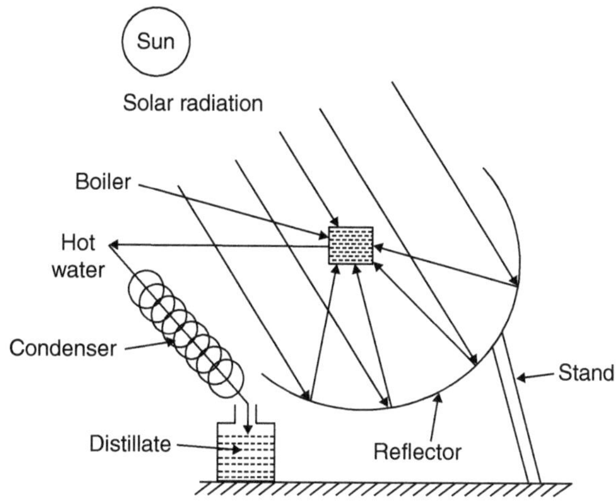

counter flow in the shell and with salted water flow of 40 l/h. Condensation occurs inside the horizontal copper tubes and it is out of film. In the same way, it works at the atmospheric pressure and it is followed by a stage of under cooling. The distillate is collected in a suitable container.

Another modified concentrating type design is proposed by Nassar et al. (2007). It is very similar to the previous design but differs with a metallic container (to resist the imposed outer pressure) of elliptical shape. It has a large absorbing surface area in the focus of the concave reflector, with a vacuum of 562.5 torr (25 kPa absolute), to reduce the normal boiling point of the feed water, and a compressor to work as a vacuum pump driven by AC power supply. The productivity of the still was reported about 20 l/day per unit area of the reflector. There was a significant improvement in the productivity of desalinized water, about 300% as compared to the other thermal solar stills. Moreover, the increase of the performance ratio is about 900% more than the greenhouse type desalination solar systems.

6.5.2.3 Solar Still Coupled with Hot Water Storage Tank Through Heat Exchanger

A solar still coupled with a hot water storage tank through heat exchanger (Voropoulos et al. 2003) is shown in the Fig. 6.20. It is a greenhouse asymmetric type, the basin of which is in direct contact with a hot water storage tank as an integral part of the still. It ensures continuous transfer of heat from the hot water tank to the basin saline water. A water storage tank is integrated below the solar still. This is done by designing the system in such way so that the bottom of the basin is actually the top of the storage tank. Heating of the tank water is done through a fin-and-tube heat exchanger placed inside the tank, using an electric heater. The heating installation is equipped with a temperature regulating device to keep the tank water temperature almost constant at the desired limits.

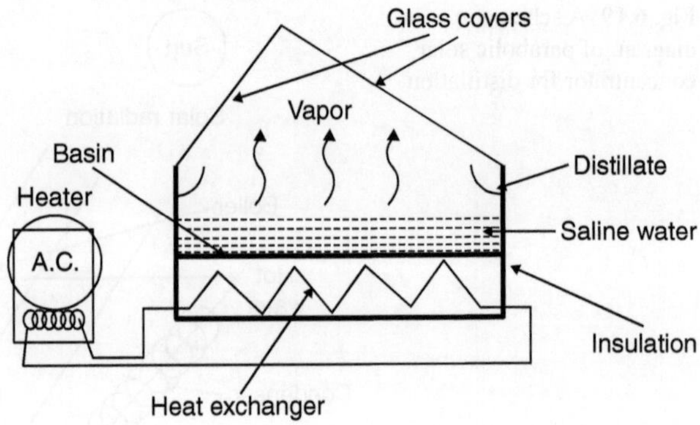

Fig. 6.20 Schematic diagram of the solar still-storage tank distillation system

6.6 Various Methods of Fixing the Glass Cover onto the Solar Still Walls

There are many methods of fixing the glass cover on the walls of a solar still, as shown in Fig. 6.21. The glass cover is fixed by using glass putty, nut and bolts, or rubber sealing to prevent the leakage of the water vapor. The glass cover can expand after heating by the solar radiation otherwise it may break.

6.7 Heat Transfer and Thermal Modeling

Thermal modeling of a solar still is a methodology to develop various theoretical equations based on various heat transfers. All heat exchanges plays an important role in the performance of the solar still that defines temperatures of basin, water, inner/outer surface of glass, and yield. It is interesting to note that on one hand the desire is to store maximum energy, and on the other hand, it is to lose maximum heat energy, for the maximum distillate production in solar stills. Figure 6.22a shows the various energy exchanges in the single slope passive solar still. Figure 6.22b shows the electrical analogy of the single slope solar still where q is the heat transfer from various components and C is the heat capacities. The electrical analogy considers various heat transfer and heat capacities at different stages such as water-glass, glass-ambient, and water-ambient.

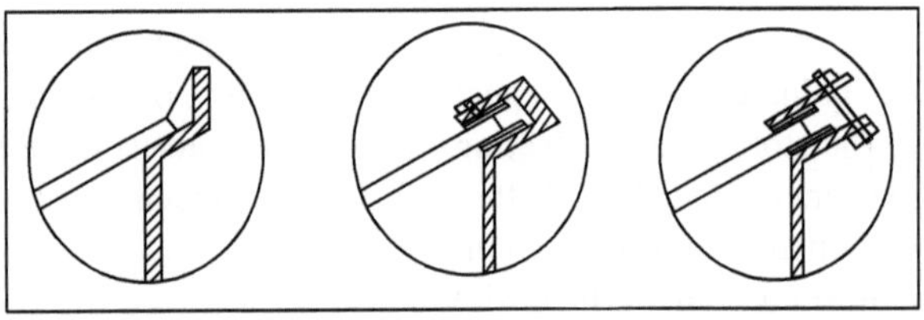

Fig. 6.21 Schematic diagram of fixing the glass cover over the solar still walls

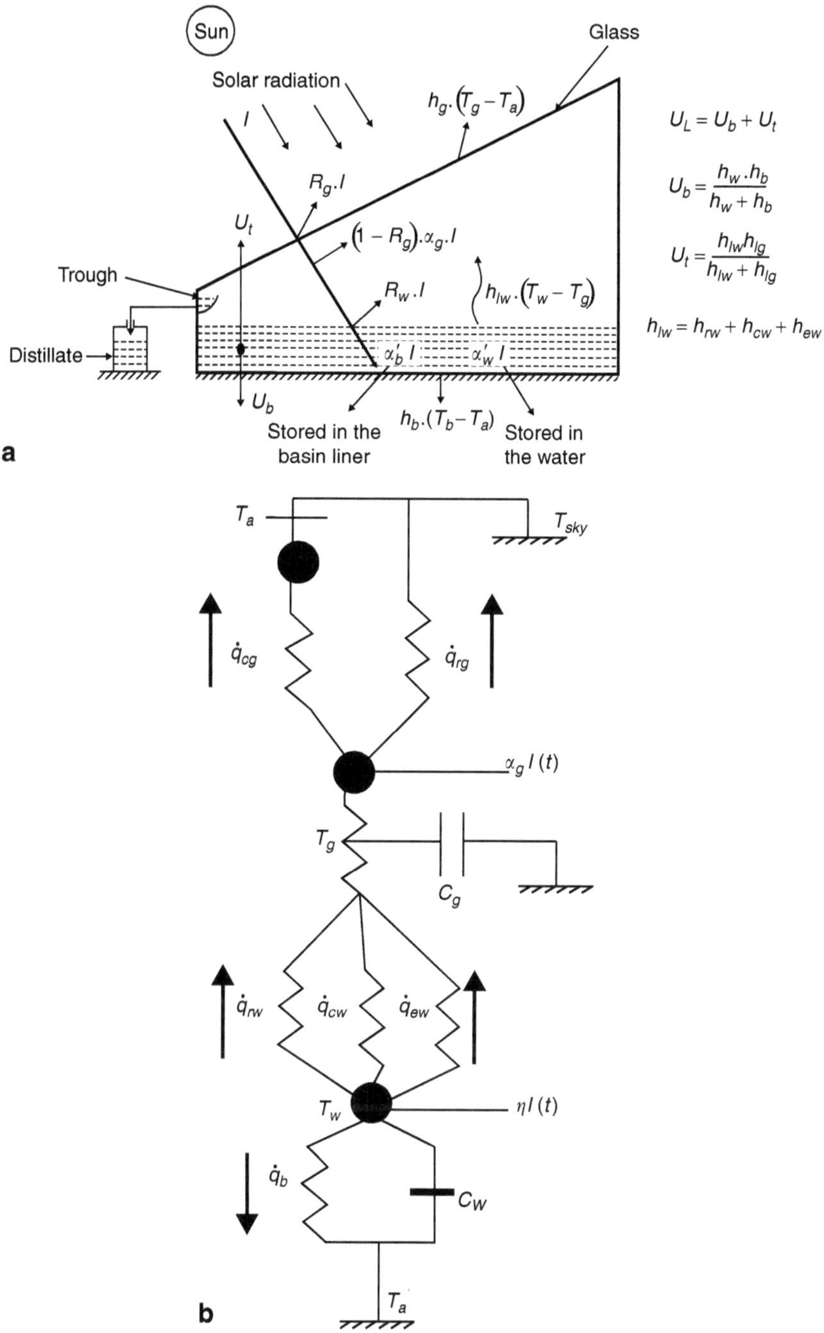

Fig. 6.22 **a** Energy flow diagram of a conventional single slope solar still, and **b** thermal circuit diagram of a conventional single slope solar still

The solar radiation, after reflection and absorption by the glass cover is transmitted inside an enclosure of the distiller unit. The transmitted radiation ($\tau_g\ I(t)$) is further partially reflected ($R'_w\ I(t)$) and absorbed ($\alpha'_w\ I(t)$) by the water mass. The attenuation of solar flux in water mass depends on its absorptivity and depth. Finally, the solar radiation reaches the blackened surface where it is mostly absorbed. Following absorption of solar radiation at the blackened surface, generally known as the basin liner, most of the thermal energy is convected to water mass and a small quantity is lost to the atmosphere by conduction. Consequently, the water is heated, leading to an increased difference of water and glass cover T temperature. There are basically three modes of heat transfer: radiation (q_{rw}), convection (q_{cw}), and evaporation (q_{ew}) from the water surface to the glass cover. The evaporated water is condensed on the inner surface of the glass cover after releasing the latent heat. Under gravity, condensed water trickles into the channels placed at the lower ends of the glass covers.

The collected water in the channel is taken out of the system for further use. The thermal energy received by the glass cover, through radiation, convection, and latent heat, is lost to the ambient by radiation and convection.

The fraction of solar flux, at different components of the distiller unit (Tiwari et al. 1989) can be mathematically expressed as:

Solar flux absorbed by the glass cover,

$$\alpha'_g = \left(1 - R_g\right)\alpha_g \cdot I(t) \tag{6.1}$$

solar flux reflected by the water mass,

$$R'_w = \left(1 - R_g\right)\left(1 - \alpha_g\right) R_w \cdot I(t) \tag{6.2}$$

solar flux absorbed by the water mass,

$$\alpha'_w = \left(1 - \alpha_g\right)\left(1 - R_g\right)(1 - R_w)\,\alpha_w \cdot I(t) \tag{6.3}$$

solar flux absorbed by the basin liner,

$$\alpha'_b = \alpha_b\left(1 - R_g\right)\left(1 - \alpha_g\right)(1 - R_w)(1 - \alpha_w) \cdot I(t) \tag{6.4}$$

solar flux lost to the ambient, through water and glass cover will be

$$(1 - \alpha_b)\left(1 - R_g\right)\left(1 - \alpha_g\right)(1 - R_w)(1 - \alpha_w) \cdot I(t) \tag{6.5}$$

where symbols R and α denotes reflectivity and absorptivity; g, w, and b are the suffix for glass, water, and basin; and $I = I(t)$ is for solar radiation.

If, however, the attenuation of solar flux within the water mass is considered, then solar flux absorbed by the water mass

$$\alpha'_w = \left(1 - \alpha_g\right)\left(1 - R_g\right)(1 - R_w) \cdot I(t)\left[1 - \sum \mu_j \exp\left(-\eta_j d_w\right)\right] \tag{6.6}$$

where, $\sum \mu_j \exp(-\eta_j d_w)$, d_w is attenuation factor and water depth.

Solar flux absorbed by the basin liner

$$\alpha_b' = \alpha_b (1 - \alpha_g)(1 - R_g)(1 - R_w) \cdot I(t) \cdot \sum \mu_j \exp(-\eta_j d_w) \quad (6.7)$$

Therefore, energy lost to the ambient, through water mass and glass cover

$$(1 - \alpha_b)(1 - \alpha_g)(1 - R_g)(1 - R_w) \cdot I(t) \cdot \sum \mu_j \exp(-\eta_j d_w) \quad (6.8)$$

The values of μ_j and η_j for different j are given in Table 6.4. The variation of attenuation factor with water depth and fraction of energy absorbed (α_w') in water mass for different absorptivity are given in Tables 6.5 and 6.6, respectively.

6.7.1 Elements of Heat Transfer

The heat transfer in solar distillation systems can be classified in terms of external and internal modes. The external heat transfer mode is primarily governed by con-

Table 6.4 The values of μ_j and η_j

j	$\mu_j \eta_j$ (m^{-1})	η_j (m^{-1})
1	0.237	0.032
2	0.193	0.45
3	0.167	3
4	0.179	35
5	0.124	255

Table 6.5 Variation of attenuation factor with water depth

d_w	$\sum \mu_j \exp(-\eta_j d_w)$
0.20	0.510
0.10	0.5492
0.08	0.5648
0.06	0.5858
0.04	0.6185
0.02	0.6756

Table 6.6 Variation of α_b' with α_w (with dye) for d_w=0.10, $R_g = R_w = 0.05$ and $\alpha_b = 0.8$

α_w	α_b' without attenuation	α_b' with attenuation
0.0	0.7220	0.3254
0.2	0.5776	0.2604
0.4	0.4332	0.1953
0.6	0.2888	0.1302
0.8	0.1444	0.0651
1.0	0.0	0.0

duction, convection, and radiation processes, which are independent of each other. These heat transfers occur outside the solar still, from the glass cover and the bottom and side insulation. Heat transfer within the solar stills is referred to as internal heat transfer that consists of radiation, convection, and evaporation. In this case, convective heat transfer occurs simultaneously with evaporative heat transfer and these two heat transfer processes are independent of radiative heat transfer.

6.7.1.1 External Heat Transfer

(a) Top Loss Heat Coefficient Due to the small thickness of the glass cover, the temperature in the glass may be assumed to be uniform. The external heat transfer, radiation and convection losses from the glass cover to the outside atmosphere q_g can be expressed as

$$\dot{q}_g = \dot{q}_{rg} + \dot{q}_{cg} \tag{6.9}$$

$$\dot{q}_{rg} = \epsilon_g \, \sigma \left[\left(T_{go} + 273 \right)^4 - \left(T_{sky} + 273 \right)^4 \right] \tag{6.10}$$

or

$$\dot{q}_{rg} = h_{rg} \left(T_{go} - T_a \right) \tag{6.11}$$

and

$$\dot{q}_{cg} = h_{cg} \left(T_{go} - T_a \right) \tag{6.12}$$

with

$$h_{rg} = \frac{\epsilon_g \, \sigma \left[\left(T_{go} + 273 \right)^4 - \left(T_{sky} + 273 \right)^4 \right]}{\left(T_{go} - T_a \right)} \tag{6.13}$$

On substituting the expressions for q_{rg} and q_{cg} in Eq. (6.9), we get

$$\dot{q}_g = h_{lg} \left(T_{go} - T_a \right) \tag{6.14}$$

where

$$h_{lg} = h_{rg} + h_{cg} \tag{6.15}$$

The empirical relation for h_{lg} can be discussed for the following conditions:

Case (i): The expression for h_{lg} is given by

$$h_{lg} = 5.7 + 3.8 \ V \tag{6.16}$$

where, V is the wind velocity (m/s). This expression includes the effect of free convection and radiation from the glass cover as discussed by Watmuff et al. (1977).

Case (ii): In this case the radiation and convection losses are to be evaluated separately, the radiative heat transfer coefficient h_{rg} can be obtained from Eq. (6.13) and the convective heat transfer coefficient h_{cg} can be obtained from

$$h_{cg} = 2.8 + 3.0\ V \tag{6.17}$$

There is, however, no significant change in the performance of the distillation system by considering h_{lg} as represented by Case (i) or Case (ii).

(b) Bottom and Side Loss Heat Coefficients Heat is also lost from the water in the basin to the ambient through the insulation and subsequently by convection and radiation from the bottom or side surface of the basin. The bottom loss coefficient U_b can be written as

$$U_b = \left[\frac{1}{h_w} + \frac{1}{K_i/L_i} + \frac{1}{h_{cb} + h_{rb}}\right]^{-1} \tag{6.18}$$

The side heat loss coefficient U_e can be approximated as

$$U_e = U_b\ A_{ss}/A_s \tag{6.19}$$

If A_{ss} is very small in comparison to A_s, for small water depth, U_e can be neglected. Here, A_{ss} is the surface area in contact with water and A_s the area of the basin of the still.

The rate of heat loss per m^2 from basin liner to ambient can be written as

$$\dot{q}_b = h_b\,(T_b - T_a) \tag{6.20}$$

$$h_b = \left[\frac{L_i}{K_i} + \frac{1}{h_{cb} + h_{rb}}\right]^{-1} \tag{6.21}$$

6.7.1.2 Internal Heat Transfer

Heat exchange from the water surface to the inner surface of the glass cover inside the distillation unit, is governed by radiation, convection, and evaporation. Various scientists have given different correlations for heat transfer coefficients. Correlations given by Dunkle (1961) are valid for a low range of temperatures (i.e., for passive solar stills) and are proven by various scientists such as Tiwari and Tiwari (2007). These correlations are given as follows.

Radiative heat transfer:

$$q_{rw} = \varepsilon_{eff} \cdot F_{12} \cdot \sigma \left[(T_w + 273)^4 - \left(T_{gi} + 273\right)^4\right]$$
$$q_{rw} = h_{rw} \cdot F_{12} \cdot \left[\left(T_w + T_{gi}\right)\right] \tag{6.22}$$

$$h_{rw} = \varepsilon_{eff} \cdot \sigma \left[(T_w + 273)^2 + (T_{gi} + 273)^2 \right] \cdot [T_w + T_{gi} + 576] \tag{6.23}$$

Where

$$\frac{1}{\varepsilon_{eff}} = \frac{1}{\varepsilon_w} + \frac{1}{\varepsilon_g} - 1 \tag{6.24}$$

Convective heat transfer:

$$\begin{aligned} q_{cw} &= 0.884 \left[T_w - T_{gi} + \frac{(P_w - P_{gi})\,(T_w + 273)}{(268.9 \times 10^3 - P_w)} \right]^{1/3} \times (T_w - T_{gi}) \\ &= h_{cw} \times (T_w - T_{gi}) \end{aligned} \tag{6.25}$$

Evaporative heat transfer:

$$q_{ew} = 16.273 \times 10^{-3} \times h_{cw} \times (P_w - P_{gi}) \tag{6.26}$$

Where,

$$h_{ew} = 0.016273 \times h_{cw} \times \left(\frac{P_w - P_{gi}}{T_w - T_{gi}} \right) \tag{6.27}$$

and

$$\begin{aligned} P_w &= \exp\left[25.317 - \left(\frac{5{,}144}{273 + T_w} \right) \right], \\ P_{ci} &= \exp\left[25.317 - \left(\frac{5{,}144}{273 + T_{gi}} \right) \right] \end{aligned} \tag{6.28}$$

$$h_{lw} = h_{cw} + h_{ew} + h_{rw} \tag{6.29}$$

6.7.2 Overall Heat Transfer

(a) Top Loss Heat Coefficient The rate of heat lost in the upward direction of a distillation system is written as

$$\dot{q}_t = U_t\,(T_w - T_a) \tag{6.30}$$

where the top loss coefficient U_t from water surface to ambient air can be

$$U_t = \left(\frac{1}{h_{lg}} + \frac{1}{h_{lw}} \right)^{-1} \tag{6.31}$$

(b) Bottom Loss Heat Coefficient The rate of heat loss through the bottom of the insulation from water to ambient air is

$$\dot{q}_{bg} = U_b\,(T_w - T_a) \tag{6.32}$$

where U_b is the bottom heat loss coefficient.

The rate of total heat lost per 1 m^2 from the water surface to the ambient through the top and the bottom of the system can be obtained by adding Eqs. (6.30) and (6.32) as

$$\dot{q}_{loss} = U_L(T_w - T_a) \tag{6.33}$$

where $U_L = U_t + U_b$ is the overall heat transfer coefficient.

6.7.3 Instantaneous Thermal Efficiency (η_i)

It is defined as the ratio of evaporative heat transfer to the total heat input to the solar still for an instant. It can be expressed by the following correlation.

$$\eta_i = \frac{\textit{Evaporative heat transfer}}{\textit{Total incident solar radiation}} = \frac{\dot{m}_{ew}L}{I(t)A_s} \times 100 \tag{6.34}$$

6.7.4 Overall Thermal Efficiency (η_o)

$$\begin{aligned}\eta_o &= \frac{(\textit{Total yield per day}) \times (\textit{Latent heat of vaporisation})}{(\textit{Total incident solar radiation in daytime})} \\ &= \frac{\sum_{1}^{24} \dot{m}_{ew} \times L}{A_s \int I(t)dt} \times 100\end{aligned} \tag{6.35}$$

6.8 Thermal Analysis: Development of Energy Balance Equations

For the thermal analysis of solar stills, three kinds of systems were taken: conventional single slope, double slope, and active solar stills. The analyses are given as follows.

6.8.1 Conventional Single Slope Solar Still

The following assumptions are made in writing the energy balance:

1. Inclination of the glass cover is very small.
2. The heat capacity of the glass cover, absorbing material, and insulation (bottom and sides) is negligible.
3. The solar distiller unit is vapor-leakage proof.
4. Inner and outer glass cover temperatures are equal.

The energy balances for different components of the still are

Glass cover:

$$\begin{bmatrix}\textit{Rate of solar}\\ \textit{energy absorbed}\end{bmatrix}+\begin{bmatrix}\textit{Rate of energy}\\ \textit{received from water}\end{bmatrix}=\begin{bmatrix}\textit{Rate of energy}\\ \textit{lost to ambient}\end{bmatrix}$$

$$\alpha'_g I\,(t) + [\dot{q}_{rw} + \dot{q}_{ew} + \dot{q}_{cw}] = \dot{q}_{rg} + \dot{q}_{cg} \tag{6.36}$$

Water mass:

$$\begin{bmatrix}\textit{Rate of solar}\\ \textit{energy}\\ \textit{absorbed}\end{bmatrix}+\begin{bmatrix}\textit{Rate of}\\ \textit{energy given}\\ \textit{by basin}\end{bmatrix}=\begin{bmatrix}\textit{Rate of}\\ \textit{energy stored}\\ \textit{in water}\end{bmatrix}+\begin{bmatrix}\textit{Rate of energy}\\ \textit{transferred to}\\ \textit{glass cover}\end{bmatrix}$$

$$\alpha'_w I\,(t) + \dot{q}_b = (MC)_w \frac{dT_w}{dt} + \dot{q}_{rw} + \dot{q}_{ew} + \dot{q}_{cw} \tag{6.37}$$

Basin liner:

$$\begin{bmatrix}\textit{Rate of}\\ \textit{solar energy}\\ \textit{absorbed}\end{bmatrix}=\begin{bmatrix}\textit{Rate of energy}\\ \textit{transferred}\\ \textit{to water}\end{bmatrix}+\begin{bmatrix}\textit{Rate of energy lost}\\ \textit{by conduction through}\\ \textit{bottom/sides}\end{bmatrix}$$

$$\alpha'_b I\,(t) = \dot{q}_b + \left[\dot{q}_{bg} + \dot{q}_g \left(A_{ss}/A_s\right)\right] \tag{6.38}$$

From the above equations, one can get

$$\frac{dT_w}{dt} + aT_w = f(t) \tag{6.39}$$

where,

$$a = \frac{U_L}{(MC)_w}; \quad f(t) = \frac{(\alpha\tau)_{eff} I(t) + U_L T_a}{(MC)_w};$$

$$(\alpha\tau)_{eff} = \alpha'_b \frac{h_w}{h_w + h_b} + \alpha'_w + \alpha'_g \frac{h_{lw}}{h_{lw} + h_{lg}}$$

$$U_L = U_b + U_t; \quad U_b = \frac{h_w h_b}{h_w + h_b}; \quad \text{and} \quad U_t = \frac{h_{lw} h_{lg}}{h_{lw} + h_{lg}}$$

The average temperature of water T_w and average glass temperature is given by,

$$\overline{T}_w = \frac{1}{t}\int_0^t T_w dt = \frac{\overline{f}(t)}{a}\left[1 - \frac{\left(1 - e^{-a\,\Delta t}\right)}{-a\,\Delta t}\right] + T_{w0}\frac{\left(1 - e^{-a\,\Delta t}\right)}{-a\,\Delta t};$$

$$\overline{T}_g = \frac{\alpha'_g I\,(t) + h_{lw}\overline{T}_w + h_{lg}\overline{T}_a}{h_{lw} + h_{lg}} \tag{6.40}$$

The rate of evaporative heat loss and the hourly yield of the solar still are given as

$$\dot{q}_{ew} = h_{ew}\left(T_w - T_g\right); \quad \text{and} \quad \dot{m}_{ew} = \frac{h_{ew}\left(T_w - T_g\right)}{L} \times 3600 \quad kg/m^2 h \tag{6.41}$$

6.8.2 *Thermal Modeling of Double Slope Solar Still*

The following assumptions are taken into consideration for writing energy balance equations for different components of a solar still (Dwivedi and Tiwari 2008).

1. Thermal capacity of the glass covers and insulating material of the wall of a solar still has been neglected.
2. There is no temperature gradient in the water inside the basin.
3. The system is under quasi-steady state condition.

Energy Balances on East Cover:

$$\begin{aligned}\alpha'_g I_E + h_{tE}(T_w - T_{giE}) - U_{EW}(T_{giE} - T_{giW}) \\ = \frac{K_g}{L_g}(T_{giE} - T_{goE}) = h_{aE}(T_{goE} - T_a)\end{aligned} \tag{6.42}$$

where, $h_{tE} = h_{cwE} + h_{ewE} + h_{rwE}$.

Energy Balances on West Cover:

$$\begin{aligned}\alpha'_g I_W + h_{tW}(T_w - T_{giW}) + U_{EW}(T_{giE} - T_{giW}) \\ = \frac{K_g}{L_g}(T_{giW} - T_{goW}) = h_{aW}(T_{goW} - T_a)\end{aligned} \tag{6.43}$$

where, $h_{tW} = h_{cwW} + h_{ewW} + h_{rwW}$

Energy Balance for Water Mass:

$$(MC_w)\frac{dT_w}{dt} = (I_E + I_W)\alpha'_w + 2U_{bw}(T_b - T_w) - h_{tE}(T_w - T_{giE}) - h_{tW}(T_w - T_{giw}) \tag{6.44}$$

Energy Balance for Basin Liner:

$$\alpha'_b(I_E + I_W) = 2U_{bw}(T_b - T_w) + 2U_{ba}(T_b - T_a) \tag{6.45}$$

From the above energy balance equations, the differential equation obtained can be,

$$\frac{dT_w}{dt} + aT_w = f(t)$$

From this differential equation, water temperature can be determined and is given by equation

$$Tw = \frac{\overline{f(t)}}{a}\left[1 - \exp(-a\Delta t)\right] + T_{w0}\exp(-a\Delta t) \tag{6.46}$$

Then, inner and outer glass cover temperatures,

$$T_{ciE} = \frac{A_1 + A_2 T_w}{p}; \qquad T_{ciW} = \frac{B_1 + B_2 T_w}{p};$$

$$T_{goE} = \frac{\frac{K_g}{L_g}T_{giE} + h_{aE}T_a}{\frac{K_g}{L_g} + h_{aE}}; \qquad T_{goW} = \frac{\frac{K_g}{L_g}T_{giW} + h_{aW}T_a}{\frac{K_g}{L_g} + h_{aW}}$$

where, $a = \frac{1}{MC_w}\left[\frac{2U_{bw}U_{ba}}{U_{bw} + U_{ba}} + \frac{(p - A_2)h_{tE}}{p} + \frac{(p - B_2)h_{tW}}{p}\right]$ and

$$f(t) = \frac{1}{MC_w}\left[\left(\alpha'_w + \frac{\alpha'_b U_{bw}}{U_{bw} + U_{ba}}\right)(I_E + I_W) + \frac{h_{tE}A_1 + h_{tw}B_1}{p} + \frac{2U_{bw}U_{ba}T_a}{U_{bW} + U_{ba}}\right] \tag{6.47}$$

The evaporative heat transfer rate from the east and west side of a double slope solar still is given by,

$$\dot{q}_{ewE} = h_{ewE}(T_w - T_{giE}); \quad \dot{q}_{ewW} = h_{ewW}(T_w - T_{giW}) \tag{6.48}$$

And, hourly yield for both sides can be calculated with the help of the following equation

$$\dot{m}_E = \frac{\dot{q}_{ewE} \times 3{,}600}{L}; \quad \dot{m}_W = \frac{\dot{q}_{ewW} \times 3{,}600}{L} \tag{6.49}$$

6.8.3 Active Single Slope Solar Still

The energy balance equations in terms of various heat transfer coefficients of an active solar still (Tiwari 1992) are as follows.

For Inner and Outer Glass Cover:

$$\alpha'_g I_{effs} + h_{lw}\left(T_w - T_{gi}\right) = \frac{K_g}{L_g}\left(T_{gi} - T_{go}\right) = h_{fg}\left(T_{go} - T_a\right) \tag{6.50}$$

Simplifying the equations,

$$T_{gi} = \frac{\alpha'_g I_{effs} + h_{lw} T_w + \frac{k_g}{L_g} T_{go}}{h_{lw} + \frac{k_g}{L_g}}; \quad T_{go} = \frac{\alpha'_b I_{effs} + h_w T_w + h_b T_a}{h_w + h_b}$$

For Basin Liner:

$$\alpha'_b\left(1 - \alpha'_g\right)\left(1 - \alpha'_w\right) I_{effs} = h_w\left(T_b - T_w\right) + h_b\left(T_b - T_a\right) \tag{6.51}$$

After simplifying the above equations one can derive,

$$\text{basin temperature } T_b = \frac{\alpha'_{-b} I_{effs} + h_w T_w + h_b T_a}{h_w + h_b} \tag{6.52}$$

Water Mass:

$$\begin{aligned} &\dot{Q}_u + \left(\alpha'_w\right)\left(1 - \alpha'_g\right) I_{effs} + h_w\left(T_b - T_w\right) \\ &\quad = (MC)_w \frac{dT_w}{dt} + h_{lw}\left(T_w - T_{gi}\right) \end{aligned} \tag{6.53}$$

Where the rate of useful energy available from the flat plate collector is given by

$$\dot{Q}_u = A_c F_R\left[(\alpha\tau)_c\right] I_c - U_{LC}\left(T_w - T_a\right) \tag{6.54}$$

In case of

(a) Concentratic collector,

$$\dot{Q}_u = F_R\left[(\alpha\tau)_c\right] I_c - U_{LC}\left(\frac{A_r}{A_a}\right)\left(T_w - T_a\right) \tag{6.55}$$

(b) Evacuated tube solar collector,

$$\dot{Q}_u = F_R\left[(\alpha\tau)_c\right] I_c - U_{LC}\left(\frac{A_L}{A_C}\right)\left(T_w - T_a\right) \tag{6.56}$$

(c) Evacuated tube heat pipe,

Substituting the values of T_{gi}, T_{go}, and T_b, simplifying, we get (Tiwari and Tiwari 2007b),

$$\frac{dT_w}{dT} + aT_w = f(t) \tag{6.57}$$

where $a = \frac{(UA)_{eff}}{(MC)_w}$; $f(t) = \frac{(IA)_{eff} + (UA)_{eff}T_a}{(MC)_w}$

The solution of Eq. (6.57) is given as

$$T_w = \frac{\overline{f}(t)}{a}\left[1 - \exp\left(-a\Delta t\right)\right] + T_{wo}\exp\left(-a\Delta t\right) \tag{6.58}$$

where T_{wo} is the temperature of basin water at $t = 0$ and $\overline{f}(t)$ the average value of $f(t)$ for the time interval between 0 and t.

Inner, outer glass and basin temperature in terms of water temperature can be calculated using Equations of T_{ci}, T_{co}, and T_b.

The daily yield is given by

$$\dot{M}_{ew} = \sum_{i=1}^{24} \dot{m}_{ew} \tag{6.59}$$

6.9 Comparison of Distillate Yield for Different Active Solar Stills

The variation of daily yield for an active solar still integrated with flat plate collector (FPC), concentrating tube collector, ETC and ETC with heat pipe is shown in Fig. 6.23. The daily total yield of a passive solar still is 0.88 kg/m^2/day (Fig. 6.23).

6.10 Effect of Various Parameters

Various parameters such as climatic, operational and design affect the performance of a solar still. Climatic parameters include solar intensity, wind velocity, ambient temperature, humidity, and sky temperature. Operational parameters include water depth, salinity, amount of dye, and mode of operation (i.e., natural or forced circulation of water); and design parameters include manufacturing material, which depends upon its availability in the local market, basin area, inclination angle of condensing cover, and thickness of insulation. Effects of some of these parameters are given as follows.

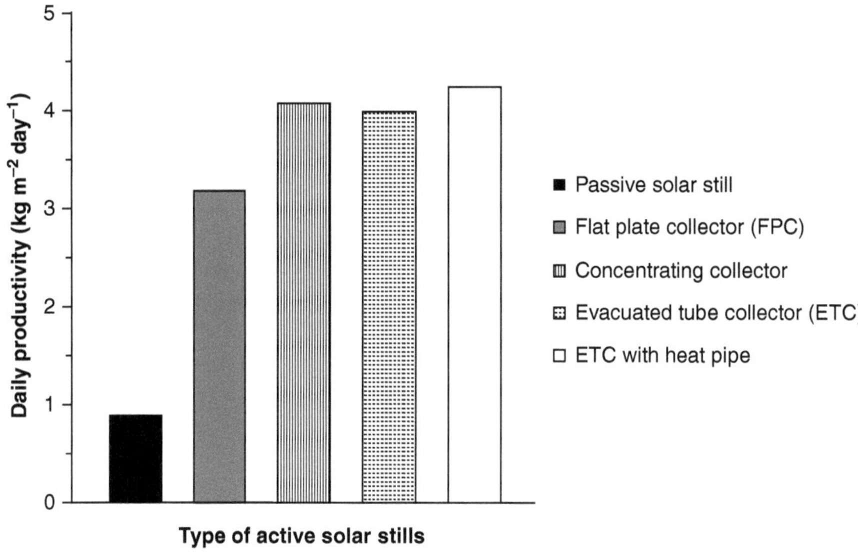

Fig. 6.23 Comparison of daily yield for active solar stills

6.10.1 Climatic Parameters

As stated previously, climatic parameters include solar intensity, wind, humidity, sky temperature, and ambient temperature. All these parameters combined together affects the solar still performance. Some of the probable cases are listed below.

1. High solar radiation, ambient temperature, but no or low wind velocity—In this kind of situation, the internal heat transfer coefficient are higher but external top heat loss reduces, which produces low yield.
2. Low solar radiation, ambient temperature, but no or low wind velocity—Because of low solar intensity, the water temperature does not rise much and hence internal heat transfer remains low, but with low wind velocity, the yield of the solar still is affected adversely.
3. High solar radiation, ambient temperature, but high wind velocity—This is a good climatic condition for high yield, due to higher internal and external top heat loss.
4. Low solar radiation, ambient temperature, but high wind velocity, etc.—This kind of condition may yield more or a similar yield in comparison to the condition stated in point 2.
5. High solar radiation, low ambient temperature—This kind of situation occurs in hilly areas such as Leh, Jammu, and Kashmir, India. It produces more yield because of high internal heat transfer and top loss.

6.10.1.1 Solar Intensity

Many scientists have found that an increase in solar intensity increases the production due to the increase in the temperature difference between water mass and glass cover. An hourly distillate output depends upon how the radiation is distributed throughout the day. Ahmed (1988) and Cooper (1983) observed a continuous increase in output from a solar still with the increase in total solar radiation. Figure 6.24 shows that in determining the output from a solar still, solar radiation is the most important parameter. The solar radiation depends upon the weather condition. In summer, the solar intensity is higher than that of winter, which results in a higher yield during the summer (Tiwari and Tiwari 2007).

6.10.1.2 Wind Velocity

For a better performance of the solar still, a low temperature of the upper surface of the glass (condensing cover) is an important parameter, which enhances condensation. The cooling of the condensing cover is done by the wind naturally. It is a concept that the higher the air flow over the condensing cover, the higher the convective heat loss to the ambient. Results of the effect of wind velocity on the performance of solar stills have been studied by various scientists. Cooper (1969) found the variation in the overall daily/annual yield between 1 and 10% because of the increase in wind velocity. Lof et al. (1961) and Yeh and Chen (1985, 1986) found a decrease in solar still production with an increase in wind velocity, whereas others observed the reverse. Morse and Read (1968) reported no significant effect of the wind over productivity. Elsherbiny and Fath (1995) observed that increasing the wind speed from 0 to 8 m/s decreases the total production by less than 10%. Toure et al. (1997) predicted that when wind speed increased from 0 to 9 m/s, the

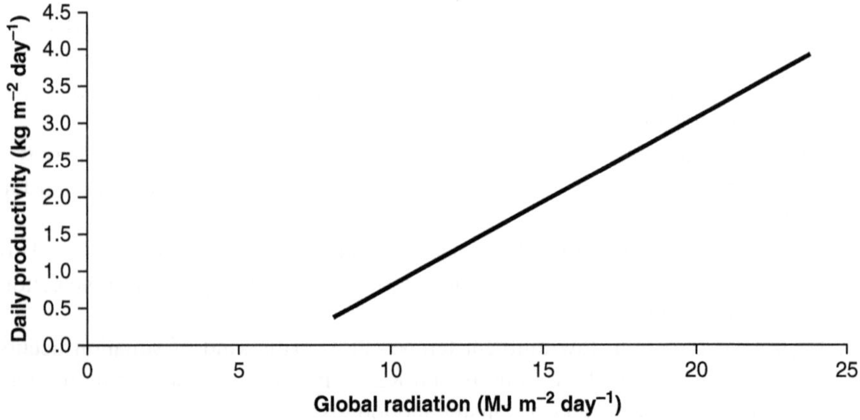

Fig. 6.24 Effect of solar radiation on rate of distillation

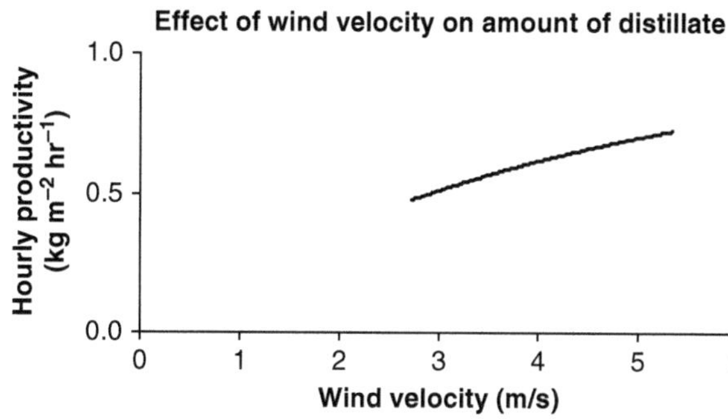

Fig. 6.25 Variation of hourly productivity with wind speed

increase in total production was less than 10%. Agrawal (1998) reported that for conventional solar stills, increasing the wind velocity from 1 to 7 m/s increased the output by 11%. According to Tiwari et al. (2007), in the case of an active solar still, if the wind velocity is increased from 1 to 8 m/s, the output increases by 45%. The results of active solar still are in accordance with those of Soliman (1976). Zurigat et al. (2004) studied the performance of regenerative solar desalination systems and they reported that if the wind speed increases from 0 to 10 m/s, the productivity can be more than 50%. Hence, one can conclude that the role of wind velocity and its effect on the productivity of the solar still is yet to be investigated thoroughly. Figure 6.25 shows the hourly variation of distillate as the wind speed varies.

6.10.1.3 Role of Sky Temperature

The effect of sky temperature on daily output is observed by taking the effect of h_{cg} and h_{rg} separately using their separate equations, respectively. It is observed that with an increase of sky temperature, total output increases in a small proportion in comparison to total output together with other climatic factors such as ambient temperature, wind velocity, and solar radiation (Tiwari et al. 2007). The sky temperature (as given in Eqs. 6.10 and 6.13 earlier in the external heat transfer section) can be obtained by any one of the following correlations,

$$T_{sky} = 0.0552 T_a^{1.5} \quad \text{(according to Swinbank 1963)}$$

$$T_{sky} = T_a - 6 \quad \text{or} \quad T_{sky} = T_a - 12\,.$$

6.10.1.4 Ambient Temperature

Ambient temperature is also an important parameter that affects a solar still performance considerably. The behavior of a solar still is different during the day compared to at night. In the day time, high ambient temperature reduces the water-glass temperature difference, which results in low yield. The solar still becomes more effective during the night with a higher water depth in the basin because of the low ambient temperature and high heat storage of water.

6.10.2 *Operational Factors*

Operational parameters are those conditions that can be varied by human beings to improve the performance of the solar still. Some of these parameters and their effects are given below.

6.10.2.1 Water Depth

Water depth not only affects the productivity of the solar still but also the design of the solar still. As water depth of the solar still increases, the height of the walls of the solar still also increases (which directly affects the cost and weight) and makes it bulkier. The heat storing capacity of water affects the solar still productivity. The larger the heat stored in the water, the more evaporation takes place. The water temperature rises rapidly when low water depth is kept in a solar still. As the water depth increases, the rise in the temperature of water slows down. As a result, the rate of evaporation and productivity decreases. This phenomenon is also called the heat loss in the solar still for the day, i.e., only the heat storage in the water mass, none or little of the production of distillate. Storage of heat leads to time delay in evaporation and hence the solar still produces distillate during the night also (i.e., in the absence of the sun). Hence, an optimized water depth is required for the entire solar still considering the design, cost, and performance. A rapid decrease in the daily/annual yield from a solar still can be up to a depth of 0.10 m and then it becomes stagnant for a passive solar still. Dev and Tiwari (2009) proved that a low water depth of 0.01 m (among 0.01, 0.08, 0.12, and 0.16 m) is best for a passive as well as an active solar still.

6.10.2.2 Effect of Material/Insulation

The type and thickness of the basin material affects the heat accumulation by the water mass of the solar still for the nocturnal operation of the solar still. A large amount of heat storage and low temperature leads to faster evaporation of the water. Nayak et al. (1980) studied the effect of increasing the thickness of the insulation on the production of distilled water. They found that productivity increases rapidly with an insulation thickness up to 4 cm; thereafter, it is affected rather slowly or at an almost constant rate (Fig. 6.26).

A study (after Tiwari and Lawrence 1991), was conducted on a double slope solar still made with different materials, i.e., fiber reinforced plastic (FRP), galvanized iron (GI), and concrete for different water depths (0.04, 0.06, 0.08, 0.10, 0.12,

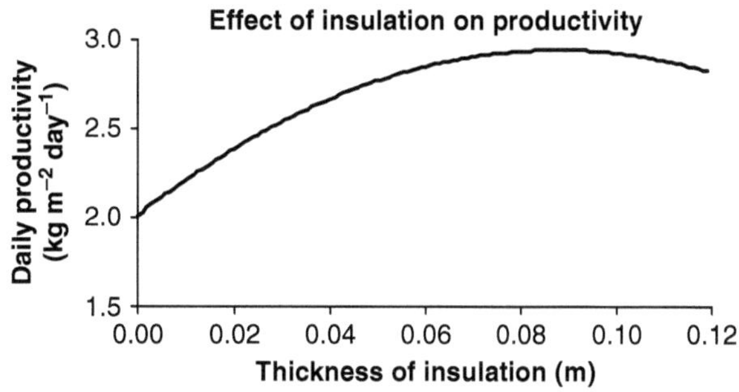

Fig. 6.26 Variation of average production per unit basin area of mounted solar stills with thickness of insulation

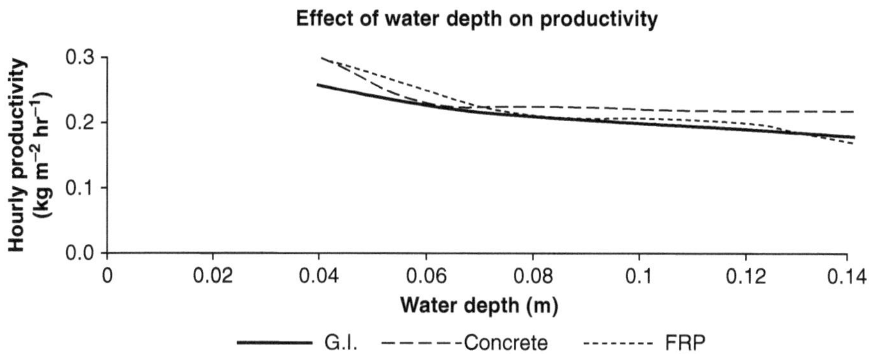

Fig. 6.27 Effect of water depth on the performance of solar still with lining/coating

and 0.14 m) in the year 1989, at the University of Papua New Guinea, Papua New Guinea (Fig. 6.27). The following were observed:

1. The performance of the FRP solar still was the same throughout the experiment.
2. The performance of the GI sheet and the concrete solar still without FRP lining is poorer than that of the FRP solar still due to the large heat loss from the sides and bottom of the still.
3. The performance of the concrete solar still is poorer than that of the GI sheet solar still because some of the energy in basin water is stored in the concrete structure.
4. FRP lining inside the surface of the concrete and GI sheet solar still reduces the heat losses from the sides and the bottom appreciably. Hence, the performance of these stills is improved significantly and approaches the performance of the FRP solar still.

6.10.2.3 Effect of Algae Formation

Algae formation is common in basins containing salt or brackish waters at temperatures up to 60°C. The absorption of solar radiation by the basin liner decreases with the presence of any solid or viscous solid object, which decreases the quantity of distillate. Hence, it requires periodic cleaning, which increases the maintenance costs.

6.10.2.4 Effect of Charcoal

Charcoal affects the performance of the solar still because of its properties: (1) wettability (fast absorption of water), (2) large absorptivity for solar radiation due to its black color, and (3) higher scattering property than reflectivity of the solar radiation. Akinsete and Aderibigbe (1983) studied the effect of charcoal pieces that enhance the performance of a solar still on partially cloudy days and during the morning hours when solar radiation remains low, and concluded that the charcoal pieces work to diffuse radiation also. However, from a maintenance point of view, using charcoal becomes a very expensive operation of the solar still because of the regular cleaning of the charcoal surface and the basin of the solar still. Wibulswas and Tadtium (1984) studied the effect of the black butyl rubber sheet and the charcoal chips (placed on water mass) on the performance of the solar still. It was reported that the performance in both cases was similar and hence the use of charcoal was recommended as economical. Based on several experiments, scientists have concluded that charcoal increases an overall efficiency by 24–28% for large water depths.

6.10.2.5 Effect of Inclination

An optimum inclination of the condensing cover for nonsymmetrical double and single-sloped stills for receiving maximum solar radiation to give the highest yield mainly depends on the location, cover material, and season. Inclination of the condensing cover varies from season-to-season. Singh and Tiwari (2004) suggested that the optimum inclination is equal to the latitude of the place. Various investigators observed that the minimum inclination of the glass cover should be at least 10° to avoid the fall of condensate back into the basin. The optimum inclination is different for the acrylic plastic cover due to the greater surface tension between the condensed water and plastic cover than that between water and glass cover. The rate of distillation per day per m^2 with different slopes of acrylic plastic covers is shown in Fig. 6.28 (Wibulswas and Tadtium 1984). The optimum inclinations of a glass cover for some locations obtained experimentally, are given in Table 6.7.

6.10.2.6 Effect of Photocatalysts

For an increment in the overall efficiency of conventional basin-type solar still, different semiconducting oxides like CuO, PbO_2, TiO_2 and MnO_2 can be used as photocatalysts (Patel et al. 2006). These oxides are good photocatalysts for chemical and pathogen degradation. To achieve this, the base of the tray is coated with

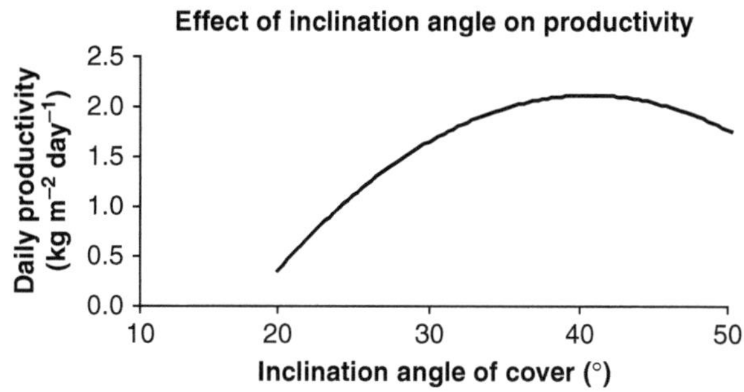

Fig. 6.28 Variation of the rate of distillation with the slope of the acrylic cover using a horizontal absorbing surface

Table 6.7 Optimum tilt of condensing cover for different locations

Location (country)	Latitude	Reference	Season	Optimum tilt (°)
Bangkok (Thailand)	13° 42′ N	Wibulswas et al. (1982)	Summer	14
Qatar	25° 30′ N	Kamal (1988)	Summer	10
Jeddah (Saudi Arabia)	21° 32′ N	Madani and Zaki (1989)	Over the year	20
Delhi (India)	28° 38′ N	Garg and Mann (1976),	Summer	15
		Malik et al. (1982),	Winter	45
		Tiwari and Tiwari (2006)	Annual	15
Tripoli (Libya)	32° 50′ N	Zaki et al. (1983)	Winter	10
Nasar (Egypt)	25°	El-Rafaie (1982)	–	30
Baghdad (Iraq)	32° N	Ahmed (1988)	Spring	30
Iraq	30° 2′ N	Abdel-Salam et al. (1986)	Summer	5
			Winter	30
Japan	40° N	Watanabe (1984)	Over the year	30
				30
Irbid (Jordan)	32° N	Tamini (1987)	Winter	45
Nigeria	10° N	Akinsete and Aderibigbe (1983)	Over the year	12

these oxides that act as a photocatalyst. A comparative analysis of the water quality from these catalyst oxides is given in Table 6.8.

6.11 Cost, Energy and Exergy Issues Related to Water Production Through Solar Stills

Solar still's cost, energy and exergy analysis are also important factors that decides its implementation worldwide. Dwivedi and Tiwari (2008) reported that the cost of the water produced by the double slope solar still, which is made of FRP, was cheaper (US$ 5.87/1,000 l) than a single slope solar still (US$ 6.92/1,000 l) when the life of the solar still is taken 50 years with constraints of proper maintenance. Some of the terms related to the cost and energy are given below (Tiwari and Tiwari 2007).

Table 6.8 Quality of water (with and without photocatalysts). (Patel et al. 2006)

Parameters analyzed	Raw saline water	Quality of distilled water			
		Without photocatalyst	With photocatalyst		
			MnO_2	PbO_2	CuO
Conductivity	0.629	0.084	0.038	0.025	0.050
TDS	700.0	368.0	48.0	52.0	48.0
Free CO_2	8.0	14.0	6.0	6.0	7.0
Total alkalinity	157.0	28.0	20.0	12.0	18.0
Chloride	42.6	14.2	14.2	14.2	14.2
Sulphate	49.4	20.6	28.8	41.2	32.9
Total hardness as $CaCO_3$	170.0	16.0	8.0	8.0	6.0
Calcium	59.7	5.5	2.5	2.5	1.7
Magnesium	5.1	0.5	0.4	0.4	0.4
Fluoride	0.5	0.2	0.3	0.3	0.1
Nitrate	4.5	1.6	2.1	1.2	1.1

All values except pH and conductivity are in ppm, conductivity is in mmhos cm^{-1}

6.11.1 Payback Period (n_p)

The following payback periods are considered for solar stills in this chapter.

6.11.1.1 Money Payback Period

The payback period is the minimum time required to recover investment cost or energy. The payback period for a system is calculated as the total investment cost divided by the annual revenues from energy saved, displaced, or produced.

$$EPBT = \frac{Total\ investment}{Annual\ revenues\ saving} \tag{6.60}$$

6.11.1.2 Energy Payback Period

It is the time required to recover the total energy spent to prepare the materials (embodied energy) for fabrication.

$$EPBT = \frac{Embodied\ energy}{Annual\ energy\ saving} \tag{6.61}$$

Tables 6.9 and 6.10 show the cost comparison and the energy analysis of the single and double slope solar stills. Energy payback time (EPBT) was reported as 1.91 and 1.85 years for single and double slope solar stills, respectively (Dwivedi and Tiwari 2008).

Table 6.9 Capital cost, salvage value and maintenance cost of the single and double slope solar still in year 2008

Components	Experimental solar still (1 US$=Rs. 47.62)	
	Single slope (US$)	Double slope (US$)
GRP body	122.8	203.21
Putty	0.63	1.26
Glass cutting	1.05	2.10
Paint	1.05	2.09
Iron stand	9.45	18.9
Gasket	2.73	5.46
Measuring cylinder	0.53	1.06
Glass cover	4.62	9.24
Capital cost (P_a)	142.86	243.32
Salvage value of iron stand after 30 years (at the rate of US$ 0.31 per kg)	6.3	12.6
Maintenance cost (Ms) (It may vary between 8 and 12% of annual capital cost)		

Table 6.10 Energy and cost analysis of single/double slope solar still considering its major components. (Dwivedi and Tiwari 2008)

Material	Energy density (kWh/kg)	Single slope	Double slope
GRP Body	25.64	9.68	24.19
GI angle iron	13.88	15	30
Glass cover	8.72	14.62	19.36
Total embodied energy (kWh)		667.47	602.73
Annual yield of basin area (kg/m^2)		465	465
Annual energy available from solar still (kWh)		350	325.5
Energy payback time (EPBT) (years)		1.91	1.85

6.12 CO_2 Emission, CO_2 Mitigation, and Carbon Credit Earned

Global warming is a major issue that is mainly caused by greenhouse gas emissions by power plants. The world is concerned about it and adopted a concept of carbon credit, which imposes a cap/limit/quota on the maximum amount of greenhouse gas emissions by country. One carbon credit is equal to one ton of carbon. Due to this cap, these countries cannot emit more carbon in the environment. If they emit less carbon, then they are able to earn more carbon for the future. This is a marketing strategy to promote environmental safety. Hence, if a renewable energy source were used instead of electricity generated from a coal based plant, a large amount of carbon can be saved, and, therefore, the environment too. The following are the analyses for solar stills that use solar energy that saves carbon to earn carbon credits.

6.12.1 CO_2 Emission

An average carbon dioxide equivalent intensity for electricity generated from coal is approximately 0.98 kg of CO_2 per kWh at source. When transmission and distribution losses for Indian conditions are considered to be 40% and domestic appliance losses are assumed to be around 20%, then the amount of CO_2 per kWh is 1.58 (Dwivedi and Tiwari 2008). Therefore,

$$\textit{Annual } CO_2 \textit{ emission} = \frac{E_{in} \times 1.58}{T_{ls}} \tag{6.62}$$

where E_{in}—Embodied energy, T_{ls}—Lifetime

and

$$\mathrm{CO_2 \text{ emission over the lifetime}} = \mathrm{E_{in}} \times 1.58 \tag{6.63}$$

6.12.2 CO_2 Mitigation: Reducing CO_2 Emission in Environment in the Form of Embodied Energy

The CO_2 mitigation (kg of CO_2) per year and over the lifetime can be expressed as,

$$\mathrm{CO_2 \text{ mitigation (kg of } CO_2) \text{ per year}} = \mathrm{E_{out}} \times 1.58 \tag{6.64}$$

and

$$\mathrm{CO_2 \text{ mitigation (kg of } CO_2) \text{ over lifetime}} = \mathrm{E_{out}} \times \mathrm{T_{LS}} \times 1.58 \tag{6.65}$$

Therefore,

$$\begin{aligned} &\text{Net } \mathrm{CO_2} \text{ mitigation (tons of } \mathrm{CO_2}) \text{ over lifetime} \\ &= (\mathrm{E_{out}} \times \mathrm{T_{LS}} - \mathrm{E_{in}}) \times 1.58 \times 10^{-3} \end{aligned} \tag{6.66}$$

The net CO_2 mitigation and emissions for the lifetime of 20 years can be evaluated by Eq. (6.64) and are given in Table 6.11.

6.12.3 Carbon Credit Earned

CO_2 has been traded at variable rates. For the study, the trade rate of US$ 22.26 per ton of CO_2 mitigation has been used. Therefore, the carbon credit earned by the system can be expressed as,

$$\text{Carbon credit earned} = (\mathrm{E_{out}} \times \mathrm{T_{LS}} - \mathrm{E_{in}}) \times 1.58 \times 10^{-3} \times 22.26. \tag{6.67}$$

Table 6.11 Estimation of CO_2 emission and mitigation for 20 year life time of single slope solar still on the basis of energy and exergy

Different Water depth (m)	0.01
Annual productivity (kg)	464.6
Input Energy E_{in} (kWh)	602.7
On the basis of energy	
Output Energy E_{out} (kWh)	325.3
CO_2 emission during the life of solar still (ton)	952.3
Net CO_2 mitigation during the life of solar still (ton)	9.3
On the basis of exergy	
Output Exergy E_{out} (kWh)	7.9
CO_2 emission during the life of solar still (ton)	952.3
Net CO_2 mitigation during the life of solar still (ton)	0

Dwivedi and Tiwari (2008) reported that the carbon credit earned on the basis of energy for a water depth of 0.01 m is US$ 207.6 and US$ 550.8, respectively for 20 and 50 years of the lifetime of a solar still.

6.13 Technology Transfer

Technology transfer is a major issue that generates the interest of a manufacturer to invest in any technology. The reliability of the solar distillation technology is high and the manufacturing process is very easy. It is also a technology that is easy to understand. But as the market demand is low due to its low performance, the technology transfer is not at a good pace. Solar distillation provides a good alternative for the production of pure water that other technologies lack in terms of input/output cost, environment protection, and ease in operation of the technology. Globally, people are now becoming more cautious about their health and environment, which is increasing the demand of the solar distillation technology. This is also increasing the interest of the manufacturer.

A technology transfer of the solar still is also associated with making distilled water drinkable, i.e., water to be used for drinking is mixed with some saline water so that the TDS levels meets the human body's mineral requirements. As the maximum desirable limit of TDS, per Indian Standard, is 1,500 ppm, the potable water to be supplied with the help of the solar still may be prepared to have similar levels of TDS. With government subsidy, there can be a further decrease in the cost of distilled water. Further, the cost of distilled water becomes cheaper with the integration of rainwater harvesting with condensing cover.

Based on the mass balance of salt in water, the following equations can be used to determine the quantity of saline water to be mixed with the demineralized water to obtain the required TDS in potable water.

$$V_b + V_d = V_w \tag{6.68}$$

and

$$S_b V_b + S_d V_d = S_w V_w \tag{6.69}$$

where V_b, V_d, and V_w are volumes of brackish, distilled, and drinking water, respectively, and S_b, S_d, and S_w are their respective salinities. The amount of saline water to be mixed with the demineralized water to obtain a TDS of 1,500 ppm can be found in Table 6.12, which shows that the quantity of saline water to be mixed decreases with the increase in TDS. It is obvious that with the decrease in the salinity of saline water, more drinking water can be prepared with low cost of potable water per liter, i.e., at a lower salinity, the solar still becomes more economical.

Technology transfer is also associated with developers of the technology (i.e., researchers and scientists). It is important to introduce and promote the technology to the users. In the case of solar distillation technology, a lot of work has been done but it is still out of reach to the common people, which is mainly because of its low productivity. The following is a hypothetical situation using solar distillation.

Let's suppose a village has a population of 5,000 people and the number of houses is 500. Each house has a solar distillation plant at its rooftop that produces 2 l of potable water per day (i.e., 24 h).

Hence,

$$\text{Total production of potable water per day} = 2 \times 500\ \text{l} = 1{,}000\ \text{l}$$

Let's suppose the salinity of brackish water, distilled water and drinking water are 5,000, 10, and 1,500 respectively.

Hence, using Eqs. (6.68) and (6.69),

$$V_b + 1{,}000 = V_w \tag{6.70}$$

and

$$5{,}000 \times V_b + 10 \times 1{,}000 = 1{,}500 \times V_w. \tag{6.71}$$

From Eqs. (6.70) and (6.71) one can calculate $V_b = 425.7$ l, $V_w = 1425.7$ l.

Table 6.12 Requirement of distilled water with varying salinity of brackish water

Brackish water		Volume of distilled water for salinity of 20 ppm V_d (l)	Volume of drinking water for salinity of 1,500 ppm V_w (l)
Salinity (ppm)	Volume V_b (l)		
Very high	0	40	40
7,000	8.53	31.47	40
6,500	9.18	30.82	40
6,000	9.95	30.05	40
5,500	10.86	29.14	40
5,000	11.94	28.06	40
4,500	13.27	26.73	40
4,000	14.94	25.06	40
3,500	17.08	22.92	40
3,000	19.93	20.07	40
2,500	23.94	16.06	40
2,000	29.95	10.05	40

On average, a man requires 6 l of water (in summer) and 3 l (in winter). So for a population of 5,000 people, 30,000 l (in summer) and 15,000 l (in winter) of water are required. In comparison, the yield produced from solar distillation is not enough to produce enough distilled water. If solar distillation is applied in combination with other technologies, then a lot of energy and money can also be saved.

If people start producing distilled water by solar distillation in every part of the world (i.e., in water efficient and deficient areas), then, water efficient areas can supply distilled water to water deficient areas. This will make a good supply and demand chain between these areas.

A double slope solar still with a provision of rain water harvesting is a better design for simultaneously promoting technology, production, money, and energy saving to the rural people. A double slope solar still made of FRP material enhances saving money due to its low cost.

On the basis of the above discussion, it is necessary to promote this technology to all places, not just in rural areas. Hence, it is also necessary to adopt good salesmanship for technology transfer.

6.14 Challenges in Adoption

Poet Samuel Taylor Coleridge (21/10/1772–25/07/1834) once wrote,

> Water, water, everywhere, Nor any drop to drink

He was not far away from the major problem of the twenty-first century, i.e., water. For solving the problem of the lack of drinking water, the world requires technology that can provide water at a fast rate, low cost, and also environmental friendly. In the past, several countries (such as India, Haiti, Mexico, Papua New Guinea, Thailand, Australia, China, Gulf countries, and USA) tried to install solar distillation plants but failed mainly due to the following reasons:

1. Lack of raw water at the site.
2. Irregular cleaning and maintenance, which decreases the output of the distillation plant because of the untrained local community handling the solar stills; although, more technical skill is not required to run a solar still plant.
3. Corrosion inside the tube, storage tank, or heat exchanger due to salty water and algae formation on the basin liner, trough, plastic tube, etc.
4. Not easy to check the level of water in the basin.
5. Breakage of glass due to sudden rain or the spilling of water on the hot glass cover when feeding, improper fitting and tightening of the glass, not providing cushion and space for thermal expansion of glass, etc.
6. Low production rate and efficiency of distilled water was not able to meet the demand.

So far, we have studied the basic principle and evolution of solar distillation, advantages/disadvantages, several passive/active still designs, thermal modeling, and CO_2 emission/carbon credit.

Over time, several designs of solar stills have been suggested. One can notice that it is difficult to choose a better solar still to fulfill all the criteria such as yield, ease of operation, cost, and mechanism. Some are good for their simplicity, such as single/double slope solar still, but their yields are low. Some are good from the point of view of yield, such as active solar stills, but their initial cost is very high. Hence, one can choose a solar still according to their need, available land area, and affordability.

From the above discussion, we can say that solar distillation is a better option for producing potable water because it uses solar energy, which is a renewable energy resource. All the nations are making efforts to harness renewable energy resources such as solar and wind. As these sources almost eliminate the possibility of degrading our environment, the use of technologies run by these resources is increasing day by day. Although, solar distillation is associated with several problems as stated earlier, it is a promising technology in the field of water purification. There are many efforts going on to eliminate these problems to implement solar distillation technology worldwide. As we have seen, solar distillation technology is a very old technology and scientists are improving it daily. For the last decade, the trend is to make this technology self-sustainable by integrating it with photovoltaic modules (PV) and using a concentrating type solar distillation system.

The water resources are fast deteriorating because of the addition of several chemicals used by human beings. Also, as the depth of underground water decreases, the possibility of several organic and inorganic chemicals (such as arsenic, phosphorus, silica, and pesticides) in the water increases. These chemicals make the water polluted and not fit to drink. There are many villages, states, and countries, where people don't have other options than to drink the polluted water. And because of that, they are facing several heath problems such as defects in teeth, osteoporosis, and kidney failure. One can obviously imagine that to remove each and every chemical to make the water fit to use, is very difficult and would require a lot of purification steps, energy, and money. Solar distillation is the only solution available so far that can effectively remove these chemicals by just one step, i.e., evaporation-condensation. For the remote areas, where electricity has not reached, solar distillation can be the only solution in comparison to other technologies such as reverse osmosis. The use of solar distillation in metropolitan cities would also help to reduce water problems as the water level is fast decreasing there. These cities face high deficiency of water in summer. Hence, the wastewater going through sewage can be reutilized by treating it with solar distillation.

In the near future, implementation of this technology would be at a higher rate because of the increasing level of problems of water at all places, concern over using renewable energy, dependability of other water treatment technologies over conventional sources of energy, and an increasing awareness over the environment.

Chapter 7
Transdisciplinary Analysis

Ravi Jain

Abstract Sustainability concepts and differing views discussed here form an important basis for transdisciplinary analysis. To effectively focus on sustainability, industrial practices need to implement technologies with due consideration to three major areas: dematerialization, rematerialization, and pollution prevention. These concepts are described and discussed in this chapter. With surprising consistency, issues of ineffective technological implementation can be traced to problems related to one of three areas: physical infrastructure, knowledge transfer, and financial resources. These issues are discussed and suggestions for addressing them are presented. Metrics for sustainability specific indicators and a framework for preference index are presented. Also discussed are elements related to technology transfer and implementation, including: characteristics of innovation and diffusion, implementation benefits, impediments, and ways to overcome impediments. In a broader sense, this chapter tries to integrate and discuss various issues related to effective use of local material, labor, and resources—focusing on economic development with due consideration to social progress, human health, and environmental protection.

Keywords Transdisciplinary analysis • Sustainability • Pollution prevention

7.1 Sustainability Concepts and Differing Views

Chapter 2 provided a comparative analysis of drinking water treatment technologies that would be considered appropriate and sustainable for developing countries and small communities in both developing and industrialized countries. This chapter included comparative studies of practices, policy reforms for implementation, and examples of successful case studies.

R. Jain (✉)
School of Engineering and Computer Science, University of the Pacific,
3601 Pacific Avenue, Stockton, CA 95211, USA
e-mail: rjain@pacific.edu

C. Ray, R. Jain (eds.), *Drinking Water Treatment,* Strategies for Sustainability,
DOI 10.1007/978-94-007-1104-4_7, © Springer Science+Business Media B.V. 2011

Focusing on goals of economic development, environmental protection, human health, and social progress, this chapter covers these issues in a broader context of sustainability and technology transfer and implementation. Some specific comments about use of local material, labor, and resources that give due consideration to economic development and social progress are represented in discussion topics that follow.

Let us review the sustainability concept as described in the Common Future (World Commission on Environment and Development 1987) and see if one can operationalize this concept in a meaningful way and provide some policy options:

> Sustainable development meets the needs of the present without compromising the ability of future generations to meet their own needs. [Sustainable development is] a process of change in which the exploitation of resources, the direction of investments, the orientation of technical development, and institutional change [are all in harmony and enhance both current and future potential to meet human needs and aspirations]. (pp. 8, 9)

Sustainable development focuses on *obligation to future generations, industrial practices, exploitation of resources,* and the role of *science and technology,* among others. Comments that follow focus on policy considerations related to some of these topics.

> If we profess to be concerned about the welfare of future generations and not terribly concerned about the welfare and needs of the poor today there is a terrible inconsistency in our thinking [stated Nobelist Robert Solow]. (Solow 1993)

We are all aware of the disparity of resources consumed by rich nations as compared to those consumed by poor nations. About one-fifth of the world's population in seven highly industrialized countries (the North) continue to consume 80% of all goods and services. Looking into the future, for every dollar of economic growth per person in developing countries (the South), a growth of $ 20 per person is expected to accrue in the North (Malone 1994). If one were to posit that large population growth would further deteriorate the environment and adversely affect sustainable development, then the situation in the South would likely get even bleaker in the future. For every person added to the population of the North, up to twenty are likely to be added to the population in the South. This growth is projected for at least the coming five decades.

This trend only further complicates the delicate balance of water, which is already over-stressed in many regions of the South (as well as the North, West, and East). By overconsumption in developed countries each individual privileged with abundant access to water is indirectly restricting access to water for those in developing regions, especially in key regions where watershed and water resources are used in common. Water, already the Earth's most precious resource, will only continue to be stressed in the future.

Within the industrialized countries, the situation of inter-generational and intragenerational equity is not very good either. Solow (1991) has pointed out that in the US, for a rich society, there is a large disparity of incomes, which includes poverty, and at the same time we do not save or invest a great deal for future generations.

One can argue in favor of the social responsibility of industry and corporations. The World Business Council for Sustainable Development (WBCSD) has given the following vague definition of Corporate Social Responsibility (CSR): "Corporate social responsibility is the commitment of business to contribute to sustainable economic development, working with employees, their families, the local community, and society at large to improve quality of life." At the World Summit on Sustainable Development (WSSD) one of the major resolutions adopted was to reduce poverty by 50% by the year 2015. This is a noble gesture, but not likely to happen for a number of reasons. One needs to understand why industry and government leaders take certain actions and make certain policy choices. Perhaps the thinking of industry is more likely to reflect what Friedman (1962) postulates regarding CSR than the noble thoughts and resolutions presented by WBCSD and WSSD.

Friedman (1962) stated that "Few trends could so thoroughly undermine the very foundations of our free society as the acceptance by corporate officials of a social responsibility other than to make as much money for their stockholders as possible. This [Corporate Social Responsibility] is a fundamentally subversive doctrine." One could argue that, notwithstanding the pretence, corporations, by and large, behave the way Friedman postulates. Therefore, by the year 2015 poverty is not going to be reduced by 50%; instead it will likely increase. In contrast to this information, there have been scattered results obtained by Yale University when plotting the Environmental Sustainability Index of each country against its Growth Competitiveness Index as compiled by the World Economic Forum (Yale University 2005; Schwab and Porter 2008). This plotted data reveals little correlation between wealth and sustainable capacity, though it cautiously indicates that a commitment to sustainability is compatible with national economic competitiveness (Yale University 2005). The hope, then, is that even developing nations can have a commitment to sustainability and still develop. The two options (development and sustainability) are not mutually exclusive. Costa Rica is an excellent example of taking extremely mature measures to protect national forests, as well as continuing to develop into the Stage 2 economy it already is (Yale University 2005). Clearly, this can be feasible for other developing countries as well.

Historically, unsustainable development and continued environmental degradation have affected the poor more than they have the rich. The poor rely heavily on common environmental resources and have few alternatives to escape the adverse consequences of pollution or environmental degradation. Examples are: urban air quality, lead paint, soil erosion, asbestos, waste treatment and disposal sites including hazardous and toxic waste disposal sites (mostly located in economically deprived areas), and groundwater contamination. In the long run, one could argue that a sustainable and healthy global industry and society are hardly compatible with a world society that is sick and is moving increasingly towards more disparity among populations, both internationally and nationally.

We continue to have difficulty in clearly understanding why these paradoxes about sustainable development exist. Most thoughtful people generally believe in the importance of sustainable development and have deeply felt views about it. Nobelist Amartya Sen wisely addressed this issue in human terms.

> Frailty of individual life and its final cessation has been well understood. This plight of the individual has been contrasted with the durability and sustainability of mankind. But now

we are seeing that the quality of life and mankind stand in danger of being further impoverished or obliterated. (Sen 2001)

7.2 Industrial Practices: Suggested Options

Industrial practices need to implement technologies with due consideration to three major areas: Dematerialization, Rematerialization, and Pollution Prevention.

Dematerialization captures efforts to reduce the physical quantity of materials required for a technology. Another aspect of dematerialization is the reduction of energy input, particularly fossil-fuel based processes. For example, China and India use four times as much energy to make a ton of steel as Japan (Roberts 1994). Building low-energy manufacturing plants is equally as important as building low-energy water treatment technologies. As an example, using solar kettles in place of flat panel collectors transitions from a technology of higher material requirement due to a large surface area, to lower material requirement as solar kettles tend to be much smaller. Additionally, manufacturing flat panel collectors and delivering those panels to project sites requires much higher specificity than developing solar kettles, as they utilize materials which have multiple uses and are available in a wider range of places.

Rematerialization reflects efforts around the globe to reuse materials in various ways. Most commonly this is done by recycling, but it can also be done through developing more biodegradable materials, or even by reducing the number of unique materials required for a technology. As an example, using aluminum-laminated cardboard for solar cooker reflectors reduces the material complexity of the system, eases waste removal, and utilizes a broadly available material to help reduce the overall impact of solar cooker technology. Using materials with "after-life demand" is also important, as it fosters a recycling mindset. In many developing economies, plastic would have a lower after-life demand than metal as it is more difficult to recycle. As an example from Chap. 2, the Chulli system uses pre-existing clay cooking structures to embed tubing into, such that the insulating material and structural support of the system is reduced to only existing components (Islam 2006). Developing technologies that can be built using local materials is another way to rematerialize: if a project is to be implemented in a region where dung, straw, and sticks are the most common building elements, then ideally the structural components of the technology could be made with those materials. This places additional stress on the development of new technologies, as the list of available and reusable materials is context specific, and creative efforts must be made to facilitate more flexible and light-weight technologies to address varying contexts, while simultaneously making technologies more accessible and more reusable to the host community.

Pollution Prevention focuses on emissions, though it has very strong ties to both dematerialization and rematerialization. While there are a few organizations committed to improving waste management ("Reuse Hawaii," http://www.reusehawaii.

org), there are no industry standards or widely accepted best management practices. The creative processes, which have begun in such organizations like "Reuse Hawaii," must continue to set precedence in all industries. By reducing the total material used, or by replacing necessary materials with biodegradable, recyclable, or readily available materials the overall pollution potential of the technology is reduced. However, each technology must still be disposed of, and each technology has some sort of emission or run-off associated with its use. Therefore, measures taken to ease disposal or operational emissions would benefit the users and the environment and foster a more sustainable methodology of production and treatment. As an example, any technology that moves from fossil fuel or electrical dependency to solar or wind energy utilization is a prime example of pollution prevention. Solar UV irradiation systems, which currently use heat lamps, could alternatively use very shallow flat-panel collectors to reduce the pollution potential of improper disposal of burnt out lamps.

A short list of concentrated efforts which will assist in the development of sustainable technology is outlined below and sorted by category, as summarized from the above paragraphs:

Dematerialization:

- Reduce materials input
- Reduce energy use in manufacturing
- Build low-energy consumption products

Rematerialization:

- Produce upgradeable, longer-lasting or biodegradable products
- Build recyclable products
- Develop technologies so that wastes are readily degradable or treatable in an environmentally safe manner
- Minimize product lifecycle environmental impact

Pollution Prevention:

- Reduce or replace hazardous or polluting materials at the source
- Use lower polluting fuels and energy technologies
- Recycle of waste material from one process as raw material for another

For each water treatment technology under consideration, the above cited parameters can be used to evaluate relative sustainability effectiveness.

Characteristically, industry has focused on products to best meet needs. As the move towards sustainable solutions continues, there must also be a shift away from products and towards services for meeting needs. The ability for manufacturers and inventors to offer education and maintenance to their products will become increasingly important as the materials used become more diverse between locations and more uniform within each location. The ability to use services to meet needs where products were used in order to retain the importance of the human variable as well as opening opportunities for local users to begin servicing products themselves will continue to become more important with time. Designing products to be localized enough to not require outside intervention allows for local servicing to become dominant after implementation. However, instead of this trend fractionalizing the world more than it already is, by removing inherent dependencies it fosters reciprocity; instead of using advanced products to meet needs it will foster a community conversation (with the "community" constantly changing definition between the geographic and technical) that relies on creative solutions through services to meet needs.

7.3 Sustainability of Technology in Developing Countries

Throughout history, many industrialized countries have developed and deployed efficient technologies and systems in developing countries. Despite overwhelming community support, many of these systems become ineffective and inoperable within a short timeframe after the implementing team pulls out. With surprising consistency this trend can be traced to one of three areas: **physical infrastructure, knowledge transfer,** and **financial resources.** The need to expand our vision to include all three of these areas is becoming increasingly clear, and many organizations are already making efforts to do so. In discussing the technical components of these technologies, it will be something that holds a lower priority to answering other challenges, but must remain a part of the equation.

7.3.1 Complexity of Physical Infrastructure

We often assume that if technology is under-supported in developing countries it is because their infrastructure is insufficient. Perhaps a different approach could be that their infrastructure is appropriate for their context and the technologies that come from developed countries are simply too complex or inappropriate. Perhaps a dependency on "safety-net" systems has developed to a point of handicapping innovative approaches. Perhaps the argument should not revolve around a lack of infrastructure in remote rural areas, but rather it should focus on inflexible designs that rely on complex infrastructure to operate.

Despite excessive intentionality on the part of scientists, engineers, and inventors to create simple technologies, often their outputs result in complexity beyond what can be supported in a developing country. Systems like the Chulli pasteurization system (Islam 2006) are as minimal as possible, and yet the problem is the dependence on burning wood or charcoal. The entire system can be explained simply: water is delivered from an elevated reservoir to a coil of pipes wrapped around a firepit and embedded in clay. Cooking pots can be placed on top of the pipe/clay structure so that cooking and pasteurization happen simultaneously. The heated water is then released from a spigot such that the rate of flow matches the rate of pasteurization. The release valve is positioned such that water escapes the system only after it has reached pasteurization temperature (Islam 2006).

As we move away from environmentally complex systems (such as the Chulli, due to the impact of burning wood), there seems to be an invariable move into technologically complex systems. Even such a simple change as running the hoses of the Chulli system through a flat panel solar pasteurizer introduces a vast realm of complexity through a single component that did not exist in earlier versions of the technology. The sensitivity of the system increases as the panel could be damaged through natural course (such as birds landing, cattle bumping, or children playing), and maintainability becomes a more challenging issue as material demands go up, and the arrangement of the collector in relation to the sun demands more technical knowledge. It also alleviates significant environmental stress, and this is the challenge: to continue to innovate systems that produce appropriate technologies recognizing social, economic, and cultural contexts.

As a technical community we must be aware of the social pressures created through the design of the technology implemented. We cannot and we will not create sustainable solutions for developing countries if we continue to innovate within our cultural context. The design of sustainable water treatment systems must be simple and honor local traditions and values. According to a UNICEF report, in order for there to be a "low" level of health concern each person must have access to a minimum of 0.1 m^3/day (26.4 gal/day) of potable water (UNICEF 2008).

As agricultural solutions come into the picture beyond humanitarian needs for survival, the demand of water increases at such a rate that the complexity is once again forcibly introduced. Factors increasing the complexity of water treatment solutions will continue to be introduced, as long as the standard of living continues to increase. Therefore, as technologies continue to come online that are simple with minimal environmental impact, there must be a sustained effort to look into the future at the impact on the quality of life, and its reciprocal impact on current treatment systems to meet increased demand.

A major part of physical infrastructure is material accessibility. While petroleum based materials are readily accessible in many of the most remote parts of the world (such as plastic bags, tarping, etc.) these products often cause environmental damage, either in their creation or in their disposal. Many factories producing parts necessary for many of these technologies are often excessively polluting and most of these materials will have to be imported from an external economy. An effort should be made to develop material for construction using the local resources and technology base to the degree possible.

Robust design of materials also becomes challenging in the more remote regions of the world. Systems need to be developed to tolerate several days' journey across uneven and unpredictable terrain, and maintain functionality throughout the cycle of seasons. Systems need to be designed such that maintenance requirements are minimal and users need to be trained to conduct regular required maintenance.

As water treatment technologies continue to mature, the need to rematerialize existing technology will continue to increase in importance. Rematerialization is simply redesigning existing technology to use recyclable/degradable, replaceable parts that are readily available. Under the mindset of rematerialization, efforts need to be taken to address material issues that have been preventing technological growth in remote regions due to inaccessible replacement parts, and will also work towards allowing maintenance to be more accessible to remote locations by severing the need for exterior input of either cash or capital.

7.3.2 *Effective Knowledge Transfer*

One of the best methods of maintainability is the education of the users for effective knowledge transfer. Perhaps it could be argued that this is the single most important component of sustainability. Even without access to material to replace components, and without the wealth to purchase any if they were available, armed with knowledge, creativity can take root and surprising solutions can be discovered with the natural resources at the disposal of the community. As stated by National Academy of Science, knowledge is the only infinite resource at the disposal of a society (NAS 2008, p. 9). As such, investment in knowledge can arguably be seen as the single most important factor in increasing the standard of living and lowering mortality rates throughout a region.

Many non-profit organizations have made education a standard element of any project of any size, and this is to be commended. For those developing new technologies, a parallel development of teaching curriculum would expedite the education process as teachers would not have to develop so much of the curriculum independently. One of the most complicating factors is continuing the cycle of education. It is quickly becoming evident that the education of the population subset that maintains the technology is no longer enough. Yet often when a project is to be developed, teaching and training occurs during the first phase of operation, but after that the inventors and implementers remove themselves from the environment to further enable and empower the local people to take ownership. This is also to be commended, but often ownership lies with those who have knowledge, and therefore it would be prudent to train teachers of teachers, as well as to encourage integration of the projects technical knowledge in the curriculum of the local secondary school to further empower the youth to take ownership not just of the project site, but more importantly of their lives and their inherent intelligence as manifested by their capacity to grasp technologically advanced concepts and ideas that provide life-saving resources for their community.

Many parts of Asia and Africa operate under a trade-apprenticeship methodology of knowledge transfer, and for those regions the same may enable water treatment systems to propagate more readily as well. If each organization that drills wells or constructs water systems of any kind were to take on one national apprentice per project, the tangible knowledge base could grow exponentially, as those apprentices could then go and develop projects under their own direction, with the financial support of their parent organization (Jain 1997, p. 201). In order to develop a true sense of knowledge and ownership, the conversations and thought-development patterns that led to a specific innovation should be made available to the users in the same manner that the physical innovation is made available. By this, users will have the capacity to expound upon and extrapolate from the current technology to arrive at new ends, which have yet to be discovered.

For these reasons it is again worth stating that investment in knowledge is the only infinite resource at the disposal of any nation or group of people, and therefore as we strive to relieve water stress for 1.1 billion people, we must also seek to alleviate "knowledge stress" (the lack of adequate knowledge) as well.

7.3.3 Adequate Financial Resources

Sustainability of implementing water treatment technologies and water supply systems is inextricably tied to adequate financial resources. While a detailed discussion of this topic is beyond the scope of this book, generally speaking, one would have to focus on items such as: development of human capital (through education and training), economic development, job creation, a stable political system, and a civil society. For industrialized nations, market-based approaches to develop revenue from water users can provide the necessary financial resources and this can also be augmented by government subsidies and tax levies. For developing countries, significant government subsidies may be needed to provide safe water simply because the poor may use unsafe water that may be available free of cost.

The development of basic economies ought to be an inherent assumption when considering the effects of drinking water treatment technology implementation. Often, communities that must travel to collect water place that responsibility on children and women; as that responsibility is alleviated through appropriate technological intervention, a larger portion of the population is available for educational and economic purposes. Therefore, as part of the technological implementation plan, there should be consideration for assisting local women in the development of basic industries such as weaving, sewing, beading, etc., to generate additional income for their families and communities.

7.4 Sustainability Framework

It is clear that thoughtful people have and will continue to have differing views about sustainability. The role of scientists and engineers is to provide information so that the policymakers can make decisions that are based on science and focus on those general goals that, in spite of varying political persuasions, provide a common ground for a great majority of humankind. For example, these goals could be:

- Achieving a reasonable standard of living for this generation without compromising the ability of future generations to meet their essential needs.
- Reducing resource exploitation to a sustainable level.

- Minimizing health and environmental impacts.
- Providing a balance between competing economic, social, and environmental needs.

If we were to generally agree on these goals, then the question arises as to how one evaluates activities designed to further this agenda and how one measures progress towards these goals. The ability to analyze different alternatives or to assess progress towards sustainability will then depend on establishing measurable entities or metrics used for sustainability. We live in an era where numbers are used to analyze different alternatives and decision making has become increasingly data-driven.

To respond to these issues, three different possibilities can be explored:

- Metrics for sustainability
- Specific indicators for sustainability
- A framework for preference index

7.4.1 Metrics for Sustainability

The book *Technological Choices for Sustainability* (Sikdar et al. 2004) provides extensive information about metrics for sustainability. The book includes numerous chapters that describe: technology sensitive indicators for sustainability, metrics for supply chain sustainability, quantifying technological aspects of process sustainability, and defining and measuring macroeconomic sustainability, etc.

The Yale Center for Environmental Law and Policy report (Yale University 2005) suggests that sustainability is a characteristic of many dynamic systems. These systems maintain themselves over time and should not be viewed as a fixed endpoint. Thus, environmental sustainability refers to the long-term maintenance of natural resources and the environment in a dynamic human context. We then have to recognize that metrics used for sustainability have to respond to the interconnectivity and temporal variations.

7.4.2 Specific Indicators for Sustainability

For developing these indicators, the Sustainability Index Report (2005) presents a comprehensive set of variables that can be helpful. This report has identified 76 indicators (or variables) grouped under five major components: environmental systems, reducing environmental stresses, reducing human vulnerability, social and institutional capacity, and global stewardship. Some examples of the indicators are: air quality, water quality, reducing ecosystem stress, natural resource management, basic human sustenance, environmental governance, private sector responsiveness, and participation in international collaborative efforts. For each of these indicators, for example water quality, there is a set of elements (e.g., TDS, suspended solids, alkalinity, and hardness) and their concentration levels provide the environmental status for that variable. These specific indicators, along with underlying set of elements, provide a comprehensive approach to analyze sustainability issues. It is important to note that in the absence of effective sustainability indicators, it is not possible for decision makers to evaluate different alternatives, policy choices, and progress towards goals.

Industrialized and developing countries have different and distinct challenges in relation to sustainability. In addition, sustainability indicators have to address issues related to economic development, economic growth, and international competitiveness. As discussed in the section on Sustainability of Technology in Developing Countries (Sect. 7.3), three major issues of concern would be: complexity of physical infrastructure, effective knowledge transfer, and adequate financial resources. For developing countries, variables under these three major elements would appear to be most important as sustainability indicators.

Thus, these indicators have to provide a general framework where, while providing means to measure progress and evaluating alternatives, they would allow industry or a nation to make choices and develop policy options. Since not all indicators and underlying elements are in common units and some elements are incommensurate, this tradeoff then naturally becomes a problem. For some cases, the target might be to reduce a damaging activity or a pollutant to a minimum level; for others, sustainability may mean striking a balance between competing priorities, and scaling variables accordingly (Yale University 2005). This is where the concept of preference index superimposed on specific indicators might be helpful.

7.4.3 A Framework for Preference Index

A model derived from the work of Keeney and Raiffa (1976), which takes into account multiple objectives, preferences, and value tradeoffs can be used to develop a framework for preference index. One of the main problems in using such an approach is the tendency on the part of some technical users to quantify items that do not lend themselves to quantification.

In developing a policy or in making specific project choices among competing demands, the decision-maker can assign utility values to consequences associated with each path instead of using explicit quantification. The payoffs are captured conceptually by associating to each path of the tree a consequence that completely describes the implications of the path. It must be emphasized that not all payoffs are in common units and many are incommensurate. This can be mathematically described as follows (Keeney and Raiffa 1976, p. 6):

$$a' \text{ is preferred to } a'' \Leftrightarrow \sum_{i=1} P_i' U_j' > \sum P_i'' U_j''$$

Where a' and a'' represent choices, P probabilities, and U utilities; the symbol $\Leftrightarrow$ reads "such that."

Utility numbers are assigned to consequences, even though some aspects of a choice are not in common units or are subjective in nature. This, then, becomes a multi-attribute value problem. This can be done informally or explicitly by mathematically formalizing the preference structure. This conceptual approach provides a generalized framework for the preference index concept and it can assist in providing a meaningful tool for including complex variables in making value tradeoffs and policy choices, and for measuring progress towards sustainability in the context of specific industry, region or a nation.

Sustainability metrics and specific indicators are useful tools; their utility in making project choices and policy decisions remain limited. Consequently, an exploration of superimposing concepts related to the preference index is suggested. This way, making project choices and making policy decisions can provide a more comprehensive framework and help address multiple objectives and competing priorities related to sustainability.

7.5 Technology Transfer and Implementation

Sustainable and clean technologies generally refer to: technologies that optimize use of resources (water, energy, land), minimize environmental impacts, produce minimum secondary wastes, and are sustainable based on current and future economic and social needs. Drinking water treatment technologies specifically discussed in the book are prime examples of sustainable and clean technologies. Thus, implementation of such technologies and associated challenges are of considerable interest from environmental, economic, and long-term societal viewpoints.

As a simple starting definition, "knowledge transfer" is a process by which knowledge is passed from one user to another. Similarly, "technology transfer" is a process by which science and technology are transferred from one individual or group to another that incorporates this new knowledge into daily life (Jain and Triandis 1997, p. 200). This definition seems intrinsic enough in the title to not be worth mentioning, except that it is inclusive of the enigma of incorporating knowledge into daily life. True technology transfer directly results in a tangible change in the daily routine of the user. In fact, knowledge transfer is merely one step towards achieving technology transfer (Jain and Triandis 1997, p. 202). The diffusion of technology occurs through different channels and involves various market agents such as private vendors, customers, consultants, and other firms, as well as public technology centers, government laboratories, and universities (Lile and Toman 1997).

Technology transfer requires sustained investment of personnel and the reiteration of knowledge as well as opportunities for technological engagement and operation by the proximal population to the project site. For a technology to be more likely to maintain a long-term presence in a social environment, it must have a team of key individuals supporting it. Those individuals can be grouped into four categories: Sponsors, Champions, Integrators, and Project Managers. Sponsors provide high-level financial backing, labor pooling, and political presence to enable the technology to move forward; Champions market the technology and solve logistical problems that occur along the way; Integrators manage conflicting priorities of inventors, locals, environment, and culture (socio-geographic constraints); and Project Managers oversee the administrative details, as well as solving technical and logistical obstacles (Jain and Triandis 1997, p. 209).

One way of evaluating a technology transfer plan would be to perform a protocol commonly used in engineering design, called *Failure Modes and Effects Analysis,* FMEA (MIL-STD-1629A 1980). In essence, FMEA recognizes that components do not always work as intended (i.e., fail), and that failure can occur in different *modes.* Furthermore, the failure mode of a component has a definite *effect* on the overall system. Failure modes may have a low or high probability of occurring (i.e., reliability), and the effect on the overall system may be anywhere from trivial to catastrophic. FMEA is a protocol that guides the designer or manager to be aware of the system's possible outcomes, particularly those modes that have a reasonable probability and a critical effect.

When exercising FMEA, the system is identified as a list of components, and the ways (modes) that each component can fail are listed, sometimes with associated probability or reliability. As an explanation of FMEA as employed in engineering design, let us contemplate a hypothetical engineering system in which one component is an electric motor. Some of the failure modes of the motor are identified as:

- Motor does not stop when turned off
- Motor does not start when switch activated
- Motor vibrates
- Motor makes excessive noise

The effect on the overall system of each of the failure modes is identified. The result may likely be an immediate mechanical failure of the system. Suitable controls and safety remediation may consequently be added to the system.

7.5.1 Characteristics of Innovation and Diffusion

The key to technology transfer is the diffusion of innovation and its absorption by various users and people groups. Five main characteristics of innovation have been outlined by Rogers (1983, 1995), as perceived by the potential adopter, which affect its rate of adoption:

- *Relative Advantage.* The degree to which the innovation is superior to ideas it supersedes.
- *Compatibility.* The degree to which the innovation is consistent with existing values, past experiences, and needs of the user.
- *Complexity.* The degree to which the innovation is relatively difficult to understand and use.

- *Trialability.* The degree to which an innovation may be tried on a limited basis (i.e., without committing to full-scale and total operational change).
- *Observability.* The degree to which the results from the use of an innovation are visible and easily communicated to users and other decision-makers.

Clearly, characteristics of innovation play an important role in technology transfer. For example, before the user adopts new technology, the user has to weigh the extra effort and investment in adopting new technology against the *relative advantages* presented by the new technology. Since existing technologies can be modified and can "stretch" to be more efficient, the new technology has to represent considerable advantages over existing ones before the extra effort involved in adopting this new technology would be considered a worthwhile undertaking.

Relative advantages relate to such items as reduced cost, increased profitability, increased convenience, reduced time, enhanced capability, and associated social status. While cost factors may stay the same or even increase, some innovations could provide relative advantage by reducing the time required to accomplish a mission or by markedly increasing product performance. For example, for military hardware such capabilities could provide a strategic or tactical advantage and thus facilitate adoption.

An innovation that is *compatible* with existing values and past experiences of the user is more likely to be adopted. For example, if the user has had a positive experience with innovations from a particular research laboratory, user adoption in the future will naturally be higher. The "felt needs" of the user can also play an important role. Sometimes external forces can create this need. For instance, regulatory requirements could create a strong need for adoption of advanced wastewater treatment technologies.

As an example, solar cooking is an innovative idea that remains foreign to much of the world's population. The introduction of solar cooking may provide significant advantages over conventional wood-burning, but this may not always be apparent to the user. Particularly if the host culture predominately cooks indoors, the idea of cooking outside may not be well accepted. Therefore, the compatibility may be a challenge, even if the advantage of not burning wood is attractive to the host community. Through education and demonstration these various perspectives may come to identify with the advantages presented within the proposed technology.

Some innovations are complex because their capabilities are difficult to understand and may require specialized training, equipment, and user capabilities. For such innovations, efforts need to be made to communicate capabilities simply and to provide the necessary training and equipment to increase the adoption rate. While a technology may be extremely efficient at answering the needs of the host community, if it is discovered to be too complex for the capacity of the community, it is then rendered impractical. Mainstream media has offered plenty of humorous examples in movies of characters developing systems that may answer a question, but are so impractical that they are laughable. In the same way, we must keep vigilant lest our best intentions become laughable to those in need of relief. Unfortunately, when our ideas are revealed to be impractical, it is far from humorous, as it is often the only recourse a region may have for pure water, and without any sort of intervention, people are left to drink contaminated water. Often it is not immediately obvious that a technology is impractical and it is therefore implemented only to be abandoned at a later date. Scalability would help pinpoint this issue before it arises.

Users are often willing to try new technology but are not willing to make full-scale and total operational changes, for obvious reasons such as uncertainty and potentially high costs. The risks outweigh the benefit to be derived. Therefore, when technology can be tried on a limited basis and if the changes can be made incrementally (*trialability*), the probability of its acceptance increases. Many innovations in office automation have followed this pattern. Where it is difficult to make a truly small-scale riverbank filtration (RBF) system (because water is being drawn from a large geologic structure, whether from one well or a hundred), it is simplistic to create a small-scale slow sand filter. Additionally, since the slow sand filter is much easier to observe and much easier to scale, it is a technology that is more likely to be accepted at higher rates than RBF. However, slow sand filtration may itself be a gateway to the acceptance of RBF. As people come to trust the Earth's ability to purify water, they may become more willing to employ the use of RBF.

If the benefits from the adoption of an innovation can be readily seen and easily communicated to potential users (*observability*), the rate of adoption is naturally greater. Computer hardware items fall into this category. Benefits from the adoption of software items (procedures, methodologies, and computer systems), however, are not as observable and not as easily communicated to potential users and thus have relatively slower rates of adoption.

Rogers' (1983) time tested framework surfaces challenges to the market diffusion of technology-based products. Marketers should consider how these factors can be incorporated into the innovation process early-on such that rapid and effective diffusion is built in as much as possible, and challenges to diffusion can be surfaced early and addressed.

7.5.2 Implementation Benefits

The implementation of Research and Development results or *Technology Transfer* accrue benefits to organizations, governmental agencies, and society. Some examples are:

- Technology transfer, in this case, can improve effectiveness and productivity of public service organizations that are responsible for providing safe drinking water to the community.
- By involving the NGOs and community leaders in implementing the technology, it can provide long-term benefits in terms of user and sponsor support.

- Specific health and social benefits from implementing appropriate and sustainable water treatment technologies are considerable. Specific rates of removal of disease-causing bacteria, protozoa, and chemicals are listed throughout the book in relation to each technology.
- Generally speaking, the economic role of new technology implementation has played (and continues to play) in the economic well-being of a nation, as documented in the National Science and Technology Council (NSTC 1999) report that states that research and development and its commercialization has enabled approximately half of the US productivity and growth in the last 50 years. The same is likely to be the case for other nations as well. Clearly, availability of safe drinking water, in addition to health benefits, will provide economic benefits as well.

Even though scientists, engineers, economists, and policy makers have long identified technology as a key input to increased productivity and creation of wealth, it is generally believed that incentives and policies for the utilization of new technology (Technology Transfer) has been lagging, thereby slowing and reducing the potential social and economic benefits. Branscomb (1993) has suggested that countries need to shift their strategies toward those centered on industry. He has further suggested that this needs to be done in a way that balances government investment in science and basic research, i.e., the creation of new knowledge, with a new focus on promoting new technology implementation centered on industry with appropriate involvement of community organizations when new technologies, such as water treatment, provide benefits that are more public than private.

The specific benefits realized will be determined on a per-use basis. In any situation where basic necessities are provided in larger quantities for less cost (social or economic) there can be an assumption that positive growth in the economic quality and quantity of the region will occur. Each of the four technologies outlined within this book, when properly implemented based on geographic, environmental, social, political, and economic factors, will provide a shift toward poverty alleviation. The size of the shift is dependent on the effectiveness of implementation, maintenance, and technology transfer.

The feasibility of implementing any of the given technologies (solar pasteurization, solar distillation, natural filtration, and membrane filtration) is inherently dependent on the source water, local materials, desired flow rate, and economic wealth of the region (or its beneficiaries). The realized gains are inherently dependent on health benefits and industrial and economic development resulting from the availability of potable water.

Speaking strictly from a perspective of resource use and allocation, there are times when it is more beneficial to use specific technologies over other comparable technologies. As an example, environments with easy access to surface water sources such as rivers or lakes, that also have a demand for a large flow rate (for large populations or large livestock populations), may benefit from building needed infrastructure to develop an RBF system. Since the flow rate produced from an RBF system running at a given optimization point is usually significantly larger than flow rates produced by other technologies' optimization points, creating a Water Board or some governing body of stakeholders may produce the accountability and cashflow to support the system.

As a contrasting example, developing a solar pasteurization system for municipal use may not be as feasible. However, solar pasteurization is much more practical than RBF if the environment provides little to no surface water, but has an abundance of sunlight. The political and financial infrastructure required to build and maintain a solar pasteurization system is also significantly reduced (when compared to RBF) for an individual household or a small community. Therefore, solar pasteurization is beneficial when the user base is small and the infrastructure is limited, while having ample access to sunlight.

Clearly, desalination may be an excellent technology to utilize if the resources available include ocean water or brackish water and exclude high rainfall, abundant sunlight, and/or large bodies of non-saline surface water.

The vast majority of locales facing a decision between multiple technologies will have choices that may point towards implementing more than one technology. As an example, if a given pasteurization technology is utilizing UV lamps as opposed to sunlight, it may require higher levels of technical competency to maintain than if the community were to use a solar still. In selecting an appropriate technology, factors may relate to economy, technical skill, resource availability, and social preference.

7.5.3 Impediments and Challenges

A new technology clearly has to have considerable relative advantage and it has to provide significant value to the customer before it is embraced by the wider user community (Jain and Triandis 1997). Even when the new technology provides a considerable relative advantage and value to the customer, its adoption and wide-scale utilization can be a challenge. Other issues related to marketing, people (perceptions, risk aversion), process (implementation requirements, adoptability and adaptability), incentives, disincentives, regulations and organizational policies all could foster or adversely affect new technology utilization. Thus, many new ideas and knowledge related to incorporating sustainable and clean technologies have a significant time lag between the generation of these ideas and their implementation.

Regardless of how useful a technology may be, its success ultimately depends on implementation, for technology must be utilized for any benefit to be accrued. Therefore, technology implementation is a crucial activity.

Considering impediments to implementing sustainable and clean technologies, it appears there would be one other fundamental issue: implementation benefits are derived not only by the implementing organization but are also accrued by society at large, thus *benefits are a property of the commons.* Some examples are: minimizing use of natural resources, using local labor and material, minimizing environmental impacts, minimizing use of energy, and using appropriate technologies that are sustainable.

Herein also lies the difficulty in developing incentives or promoting policies for sustainable technology: there is very little capacity to track the benefits on one end, as benefits are accrued by society at large. Even those outside the direct impact of the technology often reap indirect benefits: whether it is greater access to food supplies, a healthier population, or a more stable and civil society. The secondary and tertiary benefits of technological implementation ripple throughout a larger whole and promote intangibles (i.e., attributes of lifestyle) as well as tangibles such as safe water supply.

Each technology has some real limiting factor: for RBF it is access to surface water flow; for solar pasteurization and distillation it is access to abundant sunlight; and for desalination it is access to saline or brackish water. There are also other high-level constraints: RBF has a clear bias for high flow rates, desalination requires high amounts of electricity and pressure, and solar pasteurization/distillation require materials that are either fragile or difficult to manufacture. These are all extremely broad categorizations for a decision-making process, and specific benefits and constraints have been further explored in earlier chapters.

7.5.4 Ways to Overcome Impediments

Four different approaches may be considered to remove impediments and facilitate implementation of sustainable and clean technologies

- Market-based approach
- Increased investment in sustainable technologies
- Governmental subsidies
- Recognition for industry and agencies embracing and implementing such technologie

Market-based approaches can be very effective. If an agency or industry, by using clean technologies, reduces emissions at a level much lower than required by regulations, the organization should be able to bank those credits and trade them in ways that could be economically beneficial. This market-based approach seems to have worked well in other cases and can provide not only the necessary economic incentives for implementation but can help stimulate new technology developments as well. In the case of developing countries, the market-based approach could focus on providing safe water at an affordable price: a price that is less than the labor, time, and adverse health effects resulting from the use of unsafe water otherwise available. Implementation challenges always exist in cities where some free water is available to the poor. While limited regional availability of potable water may justify high costs of water in a strict supply-demand context, when dealing with resources that enable secondary benefits to exist—such as human health and access to primary education—other elements beyond simply supply and demand must be considered.

Increased *investment in new technology* can be quite beneficial for generating new concepts, ideas and systems that are economical and implementable. Increased investment by the public and industry in appropriate and sustainable technologies could further reduce the unit cost of safe water. Clearly, in meeting goals of improving human health and social progress, such investments by the governmental agencies and industry could be easily justified.

Most organizations view *governmental subsidies* negatively, except when it benefits their interests. This will be no different for proposed governmental subsidies to implement sustainable and clean technologies. According to one analysis (Bezdek and Wendling 2006), US governmental subsidies to the energy sector since 1950—primarily to the oil companies—amount to approximately $ 644 billion (in 2003 dollars). Similarly, in the US, large farm subsidies are provided for some very good economic and social reasons. In the case of sustainable and clean technologies, a credible case can be made for governmental subsidies for some sound reasons: minimization of externalities, reducing economic and social costs to future generations and protecting the environmental commons. For developing countries, community resources are very limited; thus, a strong case can be made for public investment (by governmental agencies, international aid organizations, and philanthropic groups) in providing safe drinking water that will clearly further the goals of human health protection, social progress, and economic development.

As an example, countries in the Middle East with accommodating climates for the development of extensive solar stills (see Chap. 6) may consider providing private industries with subsidies and incentives to develop solar stills Additionally, subsidizing the use of natural filtration may be beneficial to regions with high water flow, but lower sunlight availability such as Northern and Eastern Europe—much of the old Soviet Bloc. India and China also have high potential as the population centers are along many perennial rivers. By providing government incentives for technologies which are more "naturally" suited for the regional climate, overall investment in the development of new technologies that qualify for such subsidies may increase, causing the overall access to potable water to increase while driving development and implementation costs down.

Because of considerable public interest in sustainable development and the environment, identification and *recognition of industry and agencies* that adopt clean technologies can be helpful in many ways: it can be beneficial to industry directly by increasing customer base and for agencies this could help provide the needed public and legislative support. Providing recognition for industry and agencies supporting and partnering with local communities in providing safe drinking water could help to overcome some of the challenges. In fact, many of the NGOs (running small water supplies), international organizations, and large global corporations would find recognition they might receive from providing safe drinking water in developing countries to have not only public relations benefits but could also be helpful in developing markets for related products for their industry.

Implementation and transfer of sustainable drinking water treatment technologies present challenges beyond other technologies. Some of the suggestions made here—market-based approach, increased investment, subsidies, and recognition for industry and agencies—can further stimulate such technology development and its implementation.

References

Abbot CG (1938) Solar distilling apparatus. US Patent 2,141,330, 27 Dec 1938

Abdallah S, Badran O, Abu-Khader MM (2008) Performance evaluation of a modified design of a single slope solar still. Desalination 219:222–230

Abdel-Salam MS, Helwa NH, Mohamed MA, Shinkar M (1986) Effect of cover slope and inside air volume on the performance of a solar still. In: Alawi H (ed) Solar energy prospects in the Arab World. Pergamon, Oxford

Abdenacer PK, Nafila S (2007) Impact of temperature difference (water-solar collector) on solar-still global efficiency. Desalination 209:298–305

Abu-Arabi M, Zurigat Y (2005) Year round comparative study of three types of solar desalination units. Desalination 172:137–143

Abu-Arabi M, Zurigat Y, Al-Hinai H, Al-Hiddabi S (2002) Modeling and performance analysis of a solar desalination unit with double-glass cover cooling. Desalination 143:173

Abudalo RA, Bogatsu YG, Ryan JN et al (2005) Effect of ferric oxyhydroxide grain coatings on the transport of bacteriophage PRD1 and Cryptosporidium parvum oocysts in saturated porous media. Environ Sci Technol 39(17):6412–6419

Abu-Hijleh BA, Rababah HM (2003) Experimental study of a solar still with sponge cubes in basin. Energy Convers Manag 44:1411–1418

Abu-Jabal MS, Kamiya I, Narasaki Y (2001) Proving test for a solar-powered desalination system in Gaza-Palestine. Desalination 137:1–6

Adham S, Cheng RC, Vuong DX, Wattier KL (2003) Long Beach's dual-stage NF beats single-stage SWRO. Int Desalination Water Reuse 13:18–21

Adhikari RS, Kumar A, Kumar A (1990) Estimation of mass transfer rates in solar stills. Int J Energy Res 14(7):737–744

Adhikari RS, Kumar A, Sootha GD (1995) Simulation studies on a multi-stage stacked tray solar still. Sol Energy 54(5):317–325

Adiga MR, Adhikary SK, Narayanan PK, Harkare WP, Gomkale SD, Govindan KP (1987) Performance analysis of photovoltaic electrodialysis desalination plant at Tanote in Thar desert. Desalination 67:59–66

Affordable Desalination Collaboration (ADC) (2008) Affordable desalination profiles state of the art SWRO. http://www.affordabledesal.com/home/news/ADCCompletesProfileof SWRO3-28-08.pdf. Accessed 7 Feb 2009

Afgan NM, Darwish M, Carvalho MG (1999) Sustainability assessment of desalination plants for water production. Desalination 124:19–31

Agrawal S (1998) Computer based thermal modeling of advanced solar water distillation system: an experimental validation. PhD Thesis, IIT Delhi, India

Aggarwal S, Tiwari GN (1998) Convective mass transfer in double condensing chamber and conventional solar still. Desalination 115(2):181–188

Ahmadzadeh J (1978) Solar earth-water still. Sol Energy 20:387–391

C. Ray, R. Jain (eds.), *Drinking Water Treatment,* Strategies for Sustainability,
DOI 10.1007/978-94-007-1104-4,

Ahmed ST (1988) Study of single effect solar still with an internal condenser. Sol Wind Technol 5(6):637–643

Akash BA, Mohsen MS, Nayfeh W (2000) Experimental study of the basin type solar still under local climate conditions. Energy Convers Manag 41:883–890

Akhtamov RA et al (1978) Study of regenerative inclined-stepped solar still. Geliotekhnika 14(4):51–55

Akinsete VA, Aderibigbe DA (1983) Performance of the charcoal-bed roof type solar still. Solar World Congress at Perth, April 14–19

Alarcon-Padilla D-C, Garcia-Rodriguez L, Blanco-Galvez J (2007) Assessment of an absorption heat pump coupled to a multi-effect distillation unit within AQUASOL Project. Desalination 212:303–310

Alcolea A, Renard P, Mariethoz G, Bertone F (2009) Reducing the impact of a desalination plant using stochastic modeling and optimization techniques. J Hydrol 365:275–288

Al-Hayeka I, Badran OO (2004) The effect of using different designs of solar stills on water distillaton. Desalination 169(2):121–127

Al-Hinai H, Al-Nassri MS, Jurban BA (2002) Effect of climatic, design and operational parameters on the yield of a simple solar still. Energy Convers Manag 43:1639–1650

Al-Hinai H, Al-Nassri MS, Jurban BA (2002) Parametric investigation of a double-effect solar still in comparison with a single-effect solar still. Desalination 150:75–83

Al-karaghouli AA, Alnaser WE (2004) Experimental comparative study of the performances of single and double basin solar-stills. Appl Energy 77:317–325

Al-karaghouli AA, Alnaser WE (2004) Performance of single and double basin solar-stills. Appl Energy 78:347–354

Al-Obaidani S, Curcio E, Macedonio F, Di Profio G, Al-Hinai H, Drioli E (2008) Potential of membrane distillation in seawater desalination: thermal efficiency, sensitivity study, and cost estimation. J Membrane Sci 232:5–98

Al-Sahali M, Ettouney H (2007) Developments in thermal desalination processes: design, energy, and costing aspects. Desalination 214:227–240

Alsema EA, Nieuwlaar E (2000) Energy viability of photovoltaic systems. Energy Policy 28:999–1010

American Membrane Technology Association (AMTA) (2007) Membrane desalination costs. Fact Sheet 6 (FS-6), February, pp 1–2

Andreatta D (2009) The solar puddle. Solar Cookers International (SCI). http://solarcooking.org/pasteurization/puddle.htm. Accessed 5 July 2009

Annan AP (2005) GPR methods for hydrogeological studies. In: Rubin Y, Hubbard S (eds) Hydrogeophysics. Springer, Dordrecht (Water Sci Technol Libr 50:185–214)

Anonymous (1995) Strategic assessment of Florida's environment (SAFE) per capita water use. Florida State University, Institute of Science and Public Affairs, Program for Environmental Policy and Planning Systems. http://www.pepps.fsu.edu. Accessed 16 June 2009

Anonymous (2000) Tech brief: slow sand filtration. West Virginia University, National Environmental Services Center. http://www.ndwc.wvu.edu. Accessed 18 July 2009

Anonymous (2006a) Long term 2 enhanced surface water treatment rule (LT2ESWTR). US Environmental Protection Agency (USEPA). http://www.epa.gov/OGWDW/disinfection/lt2/compliance.html. Accessed 18 July 2009

Anonymous (2006b) Nanotube membranes to reduce desalination cost? Filtr Sep 43:12–13

Anonymous (2008) Dow teams up with Singapore's Public Utilities Board. Membrane Technology 2–3

Anonymous (nd) Arsenic—As. Lenntech Water Treatment & Air Purification Holding BV. http://www.lenntech.com/periodic-chart-elements/as-en.htm. Accessed 25 July 2009

Anonymous (nd) History of water treatment. Lenntech Water Treatment & Air Purification Holding BV. http://www.lenntech.com/history-water-treatment.htm. Accessed 25 July 2009

Anonymous (nd) Per capita use. Lake Oswego Water Treatment Plant. http://www.ci.oswego.or.us/wtp/Percapitadailyusage.htm. Accessed 16 June 2009

Anonymus (nd) Reuse Hawaii. http://www.reusehawaii.org. Accessed 27 June 2010

AWWARF (2001) Design of early warning and predictive source-water monitoring systems. American Water Works Association Research Foundation, Denver

Ayav PI, Atagunduz G (2007) Theoretical and experimental investigations on solar distillation of IZTECH campus area seawater. Desalination 208:169–180

Aybar HS (2006) Mathematical modeling of an inclined solar water distillation system. Desalination 190:63–70

Aybar HS, Egelioglu F, Atikol U (2005) An experimetal study on an inclined solar water distillation system. Desalination 180:285–289

Badran AA, Hamdan MA (1995) The inverted trickle solar still. Sol Energy 17:51–60

Badran AA, Assaf LM, Kayed KS, Ghaith FA, Hammash MI (2004) Simulation and experimental study for an inverted trickle solar still. Desalination 164(1):77–85

Badran AA, Ahmad AA, Salman IAE, Odat MZ (2005) A solar still augmented with a flat plate collector. Desalination 172:227–234

Badran OO (2007) Experimental study of the enhancement parameters on a single slope solar still productivity. Desalination 209:136–143

Badran OO, Al-Tahaineh HA (2005) The effect of coupling a flat plate collector on the solar still productivity. Desalination 183:137–142

Bakker M, Kelson VA, Luther KH (2005) Multilayer analytic element modelling of radial collector wells. Ground Water 43(6):926–934

Banat R, Jawied N (2008) Economic evaluation of desalination by small-scale autonomous solar-powered membrane distillation units. Desalination 220:566–573

Bapeshwar V, Tiwari GN (1984) Effect of water flow over the glass on the performance of a solar still coupled with a flat plate collector. Int J Sol Energy 2:277–288

Barber LB, Lee KE, Swackhamer DL, Schoenfuss HL (2007) Reproductive responses of male fathead minnows exposed to wastewater treatment plant effluent, effluent treated with XAD8 resin, and an environmentally relevant mixture of alkylphenol compounds. Aquat Toxicol 82(1):36–46. doi:10.1016/j.aquatox.2007.01.003

Bassam A, Hijleh KA, Mousa HA (1997) Water film cooling over the glass cover of a solar still including evaporation effects. J Energy 22(1):43–48

Bassam A, Hijleh KA, Rababah HM (2003) Experimental study of a solar still with sponge cubes in basin. Energy Convers Manag 44(9):1411–1418

Bauer C (2004) Marketing water, marketing reform: lessons from the Chilean experience. RFF, Washington

Baum VA, Bayaramov RB, Malevsky YM (1970) The solar stills in the desert. In: Proceedings of international solar energy congress, Melbourne, p 426

Baveye P, Berger P, Schijven J, Grischek T (2003) Research needs to improve the understanding of riverbank filtration for pathogenic microorganism removal. In: Ray C, Melin G, Linsky R (eds) Riverbank filtration: improving source-water quality. Kluwer, Dordrecht, pp 311–320

Bencala KE (2000) Invited commentary: hyporheic zone hydrological processes. Hydrol Process 14:2797–2798

Beyer W (1964) Determination of hydraulic conductivities of gravels and sands from grain size distribution. Wasserwirtsch Wassertech 14:165–169 (in German)

Beyer W, Banscher E (1976) Exploration methodology for bank filtration sites. Z Angew Geol 22(4):149–154 (in German)

Bezdek RH, Wendling RM (2006) The US energy subsidy scorecard. Issues in Science and Technology. The National Academies Press, Washington

Binley A, Kemna A (2005) DC resistivity and induced polarization methods. In: Rubin Y, Hubbard S (eds) Hydrogeophysics. Springer, Dordrecht (Water Sci Technol Libr 50:129–156)

Birkett J, Truby R (2007) A figure of merit for appreciating RO membrane performance. Int Desalination Assoc News 16

Bohrerova Z, Shemer H, Lantis R, Impellitteri CA, Linden KG (2008) Comparative disinfection efficiency of pulsed and continuous-wave UV irradiation technologies. Water Res 42(12):2975–2982

Borchardt D, Pusch M (2009) The ecology of the hyporheic zone of running waters. Adv Limnol 61:224

Bouchekima B (2003) A small solar desalination plant for the production of drinking water in remote arid areas of South Algeria. Desalination 159:197–204

Bouchekima B (2003) A solar desalination plant for domestic water needs in arid areas of South Algeria. Desalination 153:65–69

Bouchekima B, Gros B, Ouahes R, Diboun M (2001) The performance of the capillary film solar still installed in South Algeria. Desalination 137:31–38

Boukar M, Harmim A (2004) Parametric study of vertical solar still under desert climatic conditions. Desalination 168:21–28

Boukar M, Harmim A (2005) Performance evaluation of a one-sided vertical solar still tested in the desert of Algeria. Desalination 183:113–126

Bourouni K, Chaibi MT, Tadrist L (2001) Water desalination by humidification and dehumidification of air: state of the art. Desalination 137:167–176

Bradford S, Simunek J, Bettahar M, van Genuchten MT, Yates SR (2003) Modeling colloid attachment, straining, and exclusion in saturated porous media. Environ Sci Technol 37:2242–2250

Branscomb L (1993) Empowering technology policy. In: Branscomb L (ed) Empowering technology. MIT, Cambridge

Bray DH (1978) Manually operable reverse osmosis apparatus. US Patent 4,070,280, 24 Jan 1978

Budihardjo I, Morrison GL, Behnia M (2007) Natural circulation flow through water-in-glass evacuated tube solar collectors. Sol Energy 81(12):1460–1472

Burch JD, Thomas KE (1998) An overview of water disinfection in developing countries and the potential for solar thermal water pasteurization. NREL/TP-550-23110. National Renewable Energy Laboratory, Golden

Cadotte JE, Petersen RJ, Larson RE, Erickson EE (1980) A new thin-film composite seawater reverse osmosis membrane. Desalination 32:25–31

Caldwell TG (2006) Presentation of data for factors significant to yield from several riverbank filtration systems in the US and Europe. In: Hubbs SA (ed) Riverbank filtration hydrology: impacts on system capacity and water quality. Springer, Dordrecht

Camacho A (2003) Evaluation of the existing performance of infiltration galleries in alluvial deposits of the Parapeti River. In: Proceedings 2nd international Riverbank filtration conference, Cincinnati, 16–19 Sep 2003, p 207–213

Cappelletti GM (2002) An experiment with a plastic solar still. Desalination 142:221–227

Caslake L, Connolly D, Menon V, Duncanson C, Rojas R, Tavakoli J (2004) Disinfection of contaminated water by using solar irradiation. Appl Environ Microbiol 70(2):1145–1151

Cengel YA (1998) Heat transfer: a practical approach. McGraw-Hill, Manchester

Chafik E (2003) A new seawater desalination process using solar energy. Desalination 153:25–37

Chaibi MT (2000) An overview of solar desalination for domestic and agriculture water needs in remote arid areas. Desalination 127:119–133

Chaibi MT, Jilar T (2004) Practical alternatives to using solar desalination for irrigation in arid areas conditions. In: Proceedings of 3rd international conference on Solar Radiation and Day Lighting (Solaris 2007), IIT Delhi, India

Chaibi MT, Jilar T (2004) System design, operation and performance of roof-integrated desalination in greenhouses. Sol Energy 76:545–561

Chaouchi B, Zrelli A, Gabsi S (2006) Desalination of brackish water by means of a parabolic solar concentrator (Personal communication)

Chen Z, Hongfei Z, Kaiyan H, Chaochen M (2007) Steady-state experimental studies on a multi-effect thermal regeneration solar desalination unit with horizontal tube falling film evaporation. Desalination 207:59–79

Christiansen RC (2008) Sea asparagus can be oil feedstock. http://www.biodieselmagazine.com/article.jsp?article_id=2600. Accessed 7 Feb 2009

Ciochetti DA, Metcalf RH (1984) Pasteurization of naturally contaminated water with solar energy. Appl Environ Microbiol 47:223–228

Cipollina A, Sommariva C, Micale G (2005) Efficiency increase in thermal desalination plants by matching thermal and solar distillation: theoretical analysis. Desalination 183:127–136
Code of Federal Regulations (2005) Title 40: protection of environment, Part 141—National Primary Drinking Water Regulations
Coffey M (2008) Renewable energy: filtration and the green energy revolution. Filtr Sep 45:24–27
Constantz J, Stewart AE, Niswonger R, Sarma L (2002) Analysis of temperature profiles for investigating stream losses beneath ephemeral channels. Water Resour Res 38(12):1316–1328. doi:10.1029/2001 WR001221
Cooper PI (1969a) Digital simulation of transient solar still processes. Sol Energy 12(3):313–331
Cooper PI (1973b) Maximum efficiency of single effect solar still data. Sol Energy 15(3):205–214
Cooper PI (1983) Solar distillation: state of art and future prospects. Solar energy and Arab world. In: Proceedings of 1st Arab international Solar Energy conference, Kuwait, p 311
Cooper PI, Appleyard JA (1967) The construction and performance of a three effect wick type tilted solar still. Sun Work 12(1):4–8
Cox M, Rosenberry D, Su GW, Conlon T, Lee K, Constantz J (2002) Comparison of methods to estimate streambed seepage rates. In: Kenny JF (ed) Ground water/surface water interactions. American Water Resources Assoc., Middleburg, pp 519–523
Dash RR, Mehrotra I, Kumar P, Grischek T (2008) Lake bank filtration at Nainital, India: water-quality evaluation. Hydrogeol J 16:1089–1099
Dash RR, Bhanu Prakash EVP, Kumar P, Mehrotra I, Sandhu C, Grischek T (2010) River bank filtration in Haridwar, India: removal of turbidity, organics and bacteria. Hydrogeol J 18(4): 973–983
Dash ZV, Robinson BA, Zyvoloski GA (1997) Software requirements, design, and verification and validation for the FEHM application: a finite-element heat-and-mass transfer code. LA-13305-MS. Los Alamos Natl Lab, Los Alamos
Daughton CG, Ternes TA (1999) Pharmaceuticals and personal care products in the environment: agents of subtle change? Environ Health Perspect 107:907–937
Dayem AMA (2006) Experimental and numerical performance of a multi-effect condensation–evaporation solar water distillation system. Energy 31(14):2710–2727
Deb AK, Gupta A, Bandyopadhyay P, Biswas RK, Roy SK (1997) Appropriate technology for removal of arsenic from drinking water of rural West Bengal. Technical Paper No. 6062 presented at the 1997 AWWA Annual conference and Exposition, June 1997
Della Porta GB (1589) Magiae naturalis libri XX. Napoli
Delyannis E-E (1987) Status of solar assisted desalination: a review. Desalination 67:3–19
Delyannis E (2003) Historic background of desalination and renewable energies. Sol Energy 75(5):357–366
Derjaguin B, Landau L (1941) Theory of the stability of strongly charged lyophobic sols and of the adhesion of strongly charged particles in solution of electrolytes. Acta Physicochim URS 14:633–662
Dev R, Tiwari GN (2009) Characteristic equation of a passive solar still. Desalination 245 (1–3):246–265
Dev R, Tiwari GN (2010) Characteristic equation of hybrid (PV-T) active solar still. Desalination 254(1–3):126–137
Dev R, Abdul-Wahab SA, Tiwari GN (2011) Performance study of the inverted absorber solar still with water depth and total dissolved solid. Appl Energy 88(1):252–264
DHI-WASY GmbH (2008) FEFLOW® 5.4: finite element subsurface flow and transport simulation system. Institute for Water Resources Planning and System Research, Berlin
Dotremont C, Kregersman B, Puttemans S, Po H, Hanemaaijer JH (2008) Memstill: a near-future technology for sea water desalination. ICOM 2008, Honolulu, Hawaii, 12–18 July 2008
Drewes JE (2009) Ground water replenishment with recycled water: water quality improvements during managed aquifer recharge. Ground Water 47(4):502–505
Driscoll FG (1986) Groundwater and wells, 2nd ed. Johnson Division, St Paul, p 1089

Duff WS, Hodgson D (1999) Design and evaluation of a passive flow-through solar water pasteurization system. American Solar Energy Society conference—SOLAR 99, Portland, Maine, June 1999

Duff WS, Hodgson D (2000) Experiments with unpressurized solar water pasteurization systems. American Solar Energy Society conference—SOLAR 2000, Madison, Wisconsin, June 2000

Duff WS, Hodgson D (2005) A simple high efficiency solar water purification system. Sol Energy 79:25–32

Duffie JA, Beckman WA (1991) Solar engineering of thermal processes. Wiley, New York

Dunkle RV (1961) Solar water distillation: the roof type still and a multiple effect diffusion still. International Developments in Heat Transfer, ASME. In: Proceedings of international heat transfer, University of Colorado, Part V, p 895

Dutt DK, Rai SN, Tiwari GN (1988) Thermal modeling of high temperature distillation. Energy Convers Manag 28(2):151–159

Dwivedi VK (2009) Performance study of various designs of solar still. PhD Thesis, Centre for Energy Studies, IIT Delhi, New Delhi, India

Dwivedi VK, Tiwari GN (2008) Annual energy and exergy analysis of single and double slope passive solar stills. Appl Sci Res 3(3):225–241

Dwivedi VK, Tiwari GN (2010) Experimental validation of thermal model of a double slope active solar still under natural circulation mode. Desalination 250:49–55

Ebrahim S, Abdul-Jabar M, Bou-Hamad S, Safar M (2001) Fifteen years of R&D program in seawater desalination at KISR Part I: pretreatment technologies for RO systems. Desalination 135:141–153

Eckert P, Lamberts R, Irmscher R (2006) Impact of extreme low water level events in the River Rhine on the operation of a siphon system. Energie Wasser Praxis 12:20–23 (in German)

Eckert P, Lamberts R, Wagner C (2008) The impact of climate change on drinking water supply by riverbank filtration. Water Sci Technol 8(3):319–324. doi:10.2166/ws.2008.077 (IWA Publishing)

Edlinger R, Gomila FZ (1996) Ibiza seawater RO: special design features and operating data. Desalination 105:125–134

Elimelech M (1998) Particle deposition and aggregation: measurement, modeling, and simulation. Butterworth-Heinemann, Oxford

Elimelech M, Nagai M, Ko C-H, Ryan JN (2000) Relative insignificance of mineral grain zeta potential to colloid transport in geochemically heterogeneous porous media. Environ Sci Technol 34:2143–2148

El-Nashar AM (1990) Computer simulation of the performance of a solar desalination plant. Sol Energy 44(4):193–205

El-Rafaie ME (1982) Effect of step width and depth on the productivity of cascaded solar still. Proc 4th Int Conf for Mechanical Power Engineering, Cairo, Egypt

Elsayed MM (1986) Effects of parametric conditions on the performance of an ideal diffusion still. Appl Energy 22(3):187–203

El-Sayed YM (2007) The rising potential of competitive solar desalination. Desalination 216: 314–324

El-Sebaii A (2003) The effect of wind speed on active and passive solar stills. Energy Convers Manag 45(7–8):1187–1204

El-Sebaii AA (2005) Thermal performance of a triple-basin solar still. Desalination 174(1):23–37

El-Sebaii AA, Aboul-Enein S, El-Bialy E (2000) Single basin solar still with baffle suspended absorber. Energy Convers Manag 41(7):661–675

El-sherbiny SM, Fath HES (1995) Performance of diffusion stills under Egyptian climatic conditions. Sol Energy 16:1–12

El-Sherbiny IM, Lins RJ, Abdel-Bary EM, Harding DRK (2005) Eur Polym J 41(11):2584–2591

Environmental Resource Management (2010) Capacity analysis of Sonoma County Water Agency Wohler radial collector wells 1, 2, and 6. Sonoma County Water Agency, Santa Rosa

Esteban C, Franco J, Fasulo A (2005) Construction and performance of an assisted solar distiller. Desalination 173(3):249–255

Falvey HT, Todd CJ (1980) Concentratric tube solar still. (Malik et al. 1982)
Farid M, Hamad F (1993) Performance of a single basin solar still. Renew Energy 3(1):75–83
Fath HES, Hosny HM (2002) Thermal performance of a single-sloped basin still with an inherent built-in additional condenser. Desalination 142:19–27
Feachem R, Bradley D, Garelick H, Mara D (1983) Sanitation and disease: health aspects of excreta and wastewater management. Wiley, New York
Fedorenko AY (1991) Guidelines for minimizing entrainment and impingement of aquatic organisms at marine intakes in British Columbia. Canadian manuscript report of fisheries and aquatic sciences, Vancouver, British Columbia (2098)
Fernandez JL, Chargoy N (1990) Multi-stage, indirectly heated solar still. Sol Energy 44(4): 215–223
Fetter CW (2001) Applied hydrogeology, 4th edn. Prentice Hall, Upper Saddle River
Fischer T, Day K, Grischek T (2006) Sustainability of riverbank filtration in Dresden, Germany. In: UNESCO IHP-VI Series on Groundwater No. 13. Recharge systems for protecting and enhancing groundwater resources. Proceedings international symposium management of artificial recharge, Berlin, 11–16 Jun 2005, p 23–28
Foster ER, Eby S, Amos W (2005) Ten years of solar distillation application along the US-Mexico Border. Paper presented at the Solar World Congress, International Solar Energy Society, Orlando FL, 11 Aug 2005
Frick B (1970) Some new considerations about solar stills. Proceedings of international solar energy congress, Melbourne, p 395
Frick G, Sommerfeld J von (1973) Solar stills of inclined evaporating cloth. Sol Energy 14(4): 427–431
Friedman M (1962) Capitalism and freedom. University of Chicago Press, Chicago
Garcia-Rodriguez L (2003) Renewable energy applications in desalination. State of the Art. Sol Energy 75:381–393
Garcia-Rodriguez L, Palmero-Marrero AI, Gomez-Camacho C (2002) Comparison of solar thermal technologies for applications in seawater desalination. Desalination 142(2):135–142
Garg HP, Mann HS (1976) Effect of climatic, operational and design parameters on the year round performance of single-sloped and double-sloped solar still under Indian arid zone conditions. Sol Energy 18(2):159–163
Gasperikova E, Zhang Y, Hubbard S (2008) Using self potential and multiphase flow modeling to optimize groundwater pumping. Eos Trans AGU 89(53) (Fall Meet. Suppl, Abstract H42D-02)
Gollnitz WD, Whitteberry BL, Vogt JA (2004) Riverbank filtration: induced infiltration and groundwater quality. J Am Water Works Assoc 96(12):98–110
Gollnitz WD, Clancy JL, McEwen JB, Garner SC (2005) Riverbank filtration for IESWTR compliance. J Am Water Work Assoc 97(12):64–76
Gomkale SD (1988) Solar desalination as a means to provide Indian villages with drinking water. Desalination 69(2):171–176
Goosen MFA, Al-Hinai H, Sablani S (2001) Capacity-building strategies for desalination: activities, facilities and educational programs in Oman. Desalination 141(2):181–189
Gorman PD (2004) Spatial and temporal variability of hydraulic properties in the Russian River streambed, Central Sonoma County, California. Masters of science thesis, San Francisco State University
Goswami DY, Al-Kharabsheh S (2003) Experimental study of an innovative solar water desalination system utilizing a passive vacuum technique. Sol Energy 75(5):395–401
Greenlee LF, Lawler DF, Freeman BD, Marrot B, Moulin P (2009) Reverse osmosis desalination: water sources, technology, and today's challenges. Water Res 43:2317–2348
Grischek T (2003) Zur bewirtschaftung von Uferfiltratfassungen an der Elbe (Management of bank filtration sites along the River Elbe). PhD thesis, Department of Forestry, Geo- and Hydrosciences, Dresden Univ. of Technology (in German)
Grischek T, Ray C (2009) Bank filtration as a managed surface-groundwater interaction. Int J Water 5(2):125–139

Grischek T, Schubert J, Jasperse JL, Stowe SM, Collins MR (2007) What is the appropriate site for RBF? In: Fox P (ed) Management of aquifer recharge for sustainability. Proceedings ISMAR 6. Acacia, Phoenix, pp 466–474

Grischek T, Schoenheinz D, Sandhu C, Eckert P (2009) Sustainability of river bank filtration in Germany. In: Proceedings international conference Water, Environment, Energy and Society (WEES-2009), New Delhi, 12–16 Jan 2009, pp 1334–1339

Grischek T, Schoenheinz D, Syhre C, Saupe K (2010) Impact of decreasing water demand on bank filtration in Saxony, Germany. Drinking Water Eng Sci 3:11–20

Grote K, Hubbard S, Rubin Y (2003) Field-scale estimation of water content using ground-penetrating radar groundwave techniques. Water Resour Res 39(11):1321–1333. doi:10.1.1029/2003WRR002045

Gunkel G, Hoffmann A (2009) Bank filtration of rivers and lakes to improve the raw water quality for drinking water supply. In: Gertsen N, Sonderby L (eds) Water purification. Nova Science, Hauppauge, pp 137–169

Hajri I, Omri A, Nasrallah SB (2007) A numerical model for the simulation of double-diffusive natural convection in a triagular cavity using equal order and control volume based on the finite element method. Desalination 206(1–3):579–588

Hamieh BM, Beckman JR, Ybarra MD (2001) Brackish and seawater desalination using a 20 ft2 Dewvaporation Tower. Desalination 140(3):217–226

Hanemaaijer JH, van Medevoort J, Jansen AE, Dotremont C, van Sonsbeek E, Yuan T, De Ryck L (2006) Memstill membrane distillation: a future desalination technology. Desalination 199:175–176

Hanna M, Biby G, Miladinov V (2001) Production of microcrystalline cellulose by reactive extrusion. US Patent 6,228,213, 5 Aug 2001

Hanson A, Bates J (1999) Arsenic removal from water using manganese greensand: laboratory scale batch and column studies. Water Treatment Technology Report 41, p 100

Harbaugh AW, Banta ER, Hill MC, McDonald MG (2000) MODFLOW-2000, the US Geological Survey modular ground water model—User guide to modularization concepts and the ground water flow process. US Geological Survey Open-File Report 00-92, p 121

Harding J (1883) Apparatus for solar distillation. Proc Inst Civil Eng 73(1883):284–288

Heberer T, Feldmann D, Reddersen K, Altmann H, Zimmermann T (2002a) Production of drinking water from highly contaminated surface waters: removal of organic, inorganic, and microbial contaminants applying mobile membrane filtration units. Acta Hydroch Hydrob 30:24–33

Heberer T, Reddersen K, Mechlinski A (2002b) From municipal sewage to drinking water: fate and removal of pharmaceutical residues in the aquatic environment in urban areas. Water Sci Technol 3:81–88

Heeger D (1987) Investigations on clogging of river beds. PhD thesis, Dresden Univ. of Technology (in German)

Helal AM, Al-Malek SA, Al-Katheeri ES (2008) Economic feasibility of alternative designs of a PV-RO desalination unit for remote areas in the United Arab Emirates. Desalination 221:1–16

Hepbasli A (2008) A key review on exergetic analysis and assessment of renewable energy sources for a sustainable future. Renew Sust Energy Rev 12(3):593–661

Hermann M, Koschikowski J, Rommel M (2002) Corrosion-free solar collectors for thermally driven seawater desalination. Sol Energy 72(5):415–426

Hirschmann JR, Roefler SK (1970) Thermal inertia of solar stills and its influence on performance. In: Proceedings international solar energy congress, Melbourne, p 402

Hiscock KM, Grischek T (2002) Attenuation of groundwater pollution by bank filtration. J Hydrol 266(3–4):139–144

Hjulström F (1935) Studies of the morphological activity of rivers as illustrated by the River Fyris. Univ Uppsala Geol Inst Bull 25:221–527

Hodges CN (2004) Lessons from Africa…about the world. Illume 31:1–18

Hodges CN, Thompson TL, Riley JJ, Glenn EP (1993) Reversing the flow: water and nutrients from the Sea to the Land. AMBIO 22:483–490

Homonnay Z (2003) Use of bank filtration in Hungary. In: Ray C (ed) Riverbank filtration: understanding contaminant biogeochemistry and pathogen removal. NATO Science Series, IV. Earth and Environmental Sciences 14. Kluwer, Dordrecht, pp 221–228

Hongfei Z, Xiaoyan Z, Jing Z, Yuyuan W (2002) A group of improved heat and mass transfer correlations in solar stills. Energy Convers Manag 43(18):2469–2478

Hoppe-Jones C, Oldham G, Drewes JE (2010) Attenuation of total organic carbon and unregulated trace organic chemicals in U.S. riverbank filtration systems. Water Res 44(15):4643–4659

Howe ED (1961) Solar distillation research at the University of California. In: UN conference on New Sources of Energy, Rome, p 1

Hsieh PA, Wingle W, Healy RW (2000) VS2DI: a graphical software package for simulating fluid flow and solute or energy transport in variably saturated porous media. US Geological Survey, Water Resources Investigations Report 99-4130

Hubbard S, Rubin Y (2005) Introduction to Hydrogeophysics. In: Rubin Y, Hubbard S (eds) Hydrogeophysics. Springer, Dordrecht (Water Sci Technol Libr 50:3–7)

Hubbard SS, Peterson JE Jr, Majer EL, Zawislanski PT, Williams KH, Roberts J, Wobber F (1997) Estimation of permeable pathways and water content using tomographic radar data. Lead Edge 16:1623–1628 (Society of Engineering Geologists, November)

Hubbs SA (2006) Evaluating streambed forces impacting the capacity of riverbed filtration systems. In: Hubbs SA (ed) Riverbank filtration hydrology: impacts on system capacity and water quality. Springer, Dordrecht

Hubbs SA, Ball K, Caldwell T (2006) Riverbank filtration: an evaluation of RBF hydrology and impacts on yield. AWWA Research Foundation, Denver

Huisman L, Wood WE (1974) Slow sand filtration. WHO, Geneva, p 7

Hussam A, Munir AKM (2007) A simple and effective arsenic filter based on composite iron matrix: development and deployment studies for groundwater of Bangladesh. Environ Sci Health 42:1869–1878

Hutson S, Barber NL, Kenny JF, Linsey KS, Lumia DS, Maupin MA (2000) Estimated use of water in the United States in 2000. USGS Circular 1268, Reston, VA, p 46

Ishiguro M, Matsuura T, Detellier C (1995) Reverse osmosis separation for a montmorillonite membrane. J Membrane Sci 107:87–92

Islam MF, Johnston RB (2006) Household pasteurization of drinking water: the Chulli water-treatment system. J Health Popul Nutr 24(3):356–362 (International Centre for Diarrheal Disease Research, Bangladesh)

ITN Energy Systems Inc (2004) Photovolatic reverse osmosis desalination system. US Department of the Interior, Bureau of Reclamation, Denver, p 104

Jain RK, Triandis HC (1997) Management of research and development organizations: managing the unmanageable, 2nd edn. Wiley, New York

Jamaluddin ATM, Hassan AM, Al-Reweli A, Saeed MO, Bukhit LM, Al-Amri MM (2005) Operation of Al-Birk plant introducing beach well intake system. Saline Water Conversion Corporation, Saudi Arabia, Tr 3807/APP02005

Janarthanan B, Chandrasekaran J, Kumar S (2005) Evaporative heat loss and heat transfer for open- and closed-cycle systems of a floating tilted wick solar still. Desalination 180(1–3): 291–305

Janarthanan B, Chandrasekaran J, Kumar S (2006) Performance of floating cum tilted-wick type solar still with the effect of water flowing over the glass cover. Desalination 190(1–3):51–62

Jorgensen AJ, Nohr K, Sorensen H, Boisen F (1998) Decontamination of drinking water by direct heating in solar panels. J Appl Microbiol 85:441–447

Kalogirou SA (2001) Effect of fuel cost on the price of desalination water: a case for renewables. Desalination 138(1–3):137–144

Kamal WA (1988) A theoretical and experimental study of the basin-type solar still under the gulf climatic conditions. Sol Wind Technol 5(2):147–157

Kammourie N, Dajani TFF, Cioffi S, Rybar S (2008) SAWACO—North Obhor SWRO plant operational experience. Desalination 221:101–106

Karagiannis IC, Soldatos PG (2008) Water desalination cost literature: review and assessment. Desalination 223(1–3):448–456

Kausch O (1920) Die unmittelbare Ausnutzung der Sonnenenergie. Steinert, Weimar

Kaushika ND, Reddy KS (2000) Performance of a low cost solar paraboloidal dish steam generating system. Energy Convers Manag 41(7):713–726

Kee A (2006) The solar kettle-thermos flask: a cost effective, sustainable & renewable water pasteurization system for the developing world. Solar Cooking and Food Processing International conference, Granada, Spain, 12–16 July 2006

Keeney RL, Raiffa H (1976) Decisions with multiple objectives: preferences and value tradeoffs. Wiley, New York

Kennedy MD, Bremere I, Schippers JC (2002) Desalination technology. In: Smet J, van Wijk C (eds) Small community water supplies: technology, people & partnership. http://www.irc.nl. Accessed 7 Feb 2009

Khalifa A-JN, Al-Jubori AS, Abed MK (1999) An experimental study on modified simple solar stills. Energy Convers Manag 40(17):1835–1847

Khaydarov RR, Khaydarov RA (2007) Direct osmosis technique: new approach to wastewater reuse in Uzbekistan. In: Zaidi MK (ed) Wastewater reuse: risk assessment, decision-making, and environmental security. Springer, Dordrecht

Kiatsiriroat T, Bhattacharya SC, Wibulswas P (1986) Prediction of mass transfer rates in solar stills. Energy 11(9):881–886

Kiatsiriroat T, Bhattacharya SC, Wibulswas P (1987a) Transient simulation of vertical solar still. Energy Convers Manag 27(2):247–252

Kiatsiriroat T, Bhattacharya SC, Wibulswas P (1987b) Performance analysis of multiple effect vertical solar still with a flat plate solar collector. Solar Wind Technol 4(4):451–457

Kim JH, Kin KY, Tak TM (1987) Reverse osmosis of alginic acid-copper (II) complex membrane. Polymer (Korea) 11:254–260

Kim JY, Lee C, Cho M, Yoon J (2008) Enhanced inactivation of E. coli and MS-2 phage by silver ions combined with UV-A and visible light irradiation. Water Res 42(1–2):356–362

Klein G, Krebs M, Hall V, O'Brien T, Blevins BB (2005) California's water–energy relationship—Final staff report. California Energy Commission. http://www.energy.ca.gov/2005publications/CEC-700-2005-011/CEC-700-2005-011-SF.PDF

Kobayashi M (1963) A method of obtaining water in arid lands. Sol Energy 7(3):93–99

Kolpin DW, Furlong ET, Meyer MT, Thurman EM, Zaugg SD, Barber LB, Buxton HT (2002) Pharmaceuticals, hormones, and other organic wastewater contaminants in US streams, 1999–2000: a national reconnaissance. Environ Sci Technol 36(6):1202–1211

Korngold E (1982) Electrodialysis unit: optimization and calculation of energy requirement. Desalination 40:171–179

Koschikowski J, Wieghaus M, Matthias R (2003) Solar thermal-driven desalination plants based on membrane distillation. Desalination 156(1–2):295–304

Kraus KA, Shor AJ, Johnson JS (1967) Hyperfiltration studies X. Hyperfiltration with dynamically formed membranes. Desalination 2:243–266

Krauter S, Ruther R (2004) Considerations for the calculations of green house gas reduction by photovoltaic solar energy. Renew Energy 29:345–355

Kresic N (1997) Quantitative solutions in hydrogeology and groundwater modeling. CRC, New York

Krueger M, Nitzsche I (2003) The 100-year flood of the Elbe Rive in 2002 and its effects on riverbank-filtration sites. In: Melin G (ed) Riverbank filtration: the future is now! Proceedings 2nd international riverbank filtration conference, Cincinnati, 16–19 Sept 2003, p 81–85

Kruseman GP, de Ridder NA (1991) Analysis and evaluation of pumping test data, 2nd edn. Int Institute of Land Reclamation and Improvement, Wageningen

Kudish AI, Evseev EG, Walter G, Priebe T (2003) Simulation study on a solar desalination system utilizing an evaporator/condenser chamber. Energy Convers Manag 44(10):1653–1670

Kudish AI, Gale J, Zarmi Y (1982) A low cost design solar desalination unit. Energy Convers Manag 22(3):269–274

Kumar A, Anand JD (1992) Modelling and performance of a tubular multiwick solar still. Energy 17(11):1067–1071

Kumar A, Tiwari GN (1990) Use of hot water in double slope solar still through heat exchanger. Energy Convers Manag 30(2):81–89

Kumar S (2008) Performance analysis of hybrid photovoltaic/thermal (PV/T) active solar distillation system. PhD thesis, IIT Delhi, New Delhi

Kumar S, Tiwari GN (1996a) Performance evaluation of an active solar distillation system. Energy 21(9):805–808

Kumar S, Tiwari GN (1996b) Estimation of convective mass transfer in solar distillation systems. Sol Energy 57(6):459–464

Kumar S, Tiwari GN, Singh HN (2000) Annual performance of an active solar distillation system. Desalination 127(1):79–88

Kunze H (2001) A new approach to solar desalination for small- and medium-size use in remote areas. Desalination 139(1–3):35–41

Lattemann S, Hopner T (2008) Impacts of seawater desalination plants on the marine environment of the Gulf. In: Abuzinada AH, Barth HJ, Boer KB, Al Abdessalaam TZ (eds) Protecting the Gulf's marine ecosystems from pollution. Birkhauser, Basel

Lavoisier AL (1862) Oeuvres de Lavoisier, son Excellence le ministre de l'instruction publique et de cultes. 3, Takle 9

Lawand TA, Alward R (1985) A community solar still in Haiti. Intersol 85 2:1346–1350

Lawrence S, Gupta SP, Tiwari GN (1988) Experimental validation of thermal analysis of solar still with dye. Int J Sustain Energy 6(5):291–305

Lawrence SA, Tiwari GN (1990) Theoretical evaluation of solar distillation under natural circulation with heat exchanger. Energy Convers Manag 30(3):205–213

Lenntech (2009) Arsenic—As. http://www.lenntech.com/periodic/elements/as.htm

Liangxong L, Whitworth TM, Lee R (2003) Separation of inorganic solutes from oil-field produced water using a compacted bentonite membrane. J Membrane Sci 217:215–225

Libelo EL, MacIntyre WG, Seitz RD, Libelo LF (1994) Cycling of water through the sediment-water interface by passive ventilation of relict biological structures. Mar Geol 120:1–12

Lile R, Toman M (1997) Promoting international transfer of clean technology. Mutually beneficial opportunities for technology transfer to promote international greenhouse gas reductions workshop, Beijing, China, 14–17 Nov

Lindblom J, Nordell B (2007) Underground condensation of humid air for drinking water production and subsurface irrigation. Desalination 203(1–3):417–434

Lobo PC, Araujo SRD (1977) Design a simple multi-effect basin type solar still. In: Proceedings international solar energy congress, New Delhi, p 2026

Lof GOG, Eibling JA, Bloemer JW (1961) Energy balances in solar distillation. Am Inst Chem Eng 7(4):641–649

Lunt I, Hubbard S, Rubin Y (2005) Soil moisture content estimating using ground-penetrating radar reflection data. J Hydrol 307(1–4):254–269. doi:10.1016/j.jhydrol. 2004.10.014

Macler BA, Merkle JC (2000) Current knowledge on groundwater microbial pathogens and their control. Hydrogeol J 8(1):29–40

Madani AA, Zaki GM (1989) Prospective of two small water producing units. Desalination 73:167–180

Madhuri, Tiwari GN (1985) Performance of solar still with intermittent flow of waste hot water in the basin. Desalination 52:345–357

Malik MAS, Tran VV (1973) A simplified mathematical model for predicting the nocturnal output of a solar still. Sol Energy 14(4):371–385

Malik MAS, Tiwari GN, Kumar A, Sodha MS (1982) Solar distillation. Pergamon, Oxford

Malone TR (1994) The world after Rio. In: Ellis MD (ed) The role of engineering in sustainable development. American Association of Engineering Societies, Washington, pp 44–61

Mamlook R, Badran O (2007) Fuzzy sets implementation for the evaluation of factors affecting solar still production. Desalination 203(1–3):394–402

Mamtaz R, Bache DH (2001) Low-cost technique of arsenic removal from water and its removal mechanism. Technologies for Arsenic Removal from Drinking Water, p 43–58

Mani A (1980) Hand book of solar radiation data for India. Indian Allied Publishers Ltd, New Delhi

Manjikian S (1978) Reverse osmosis system adaptable for manual operation. US Patent 3,830,372

Manolakos D, Mohammed ES, Karagiannis I, Papadakis G (2008) Technical and economical comparisons between PV-RO system and RO-Solar Rankine system. Case study: Thirasia Island. Desalination 221:37–46

Mari EG, Colomer RPG, Blaise-Ombrecht CA (2007) Performance analysis of a solar still intergrated in a greenhouse. Desalination 203(1–3):435–443

Marks JW (2009) Giardiasis. MedicineNet. http://www.medicinenet.com/giardia_lamblia/ article.htm. Accessed 10 Aug 2009

Massmann G, Dünnbier U, Heberer T, Taute T (2008) Behavior and redox sensitivity of pharmaceutical residues during bank filtration—investigation of residues of phenazone-type analgesics. Chemosphere 71:1476–1485. http://dx.doi.org/doi:10.1016/ j.chemosphere.2007.12.017

McCutcheon JR, McGinnis RL, Elimelech M (2005) A novel ammonia-carbon dioxide forward (direct) osmosis desalination process. Desalination 174:1–11

McGinnis RL, Elimelech M (2007) Energy requirements of ammonia-carbon dioxide forward osmosis desalination. Desalination 207:370–382

McLaren JB, Tuttle LR (2000) Fish survival on fine mesh traveling screens. Environ Sci Policy 3:369–376

Medema GJ, Juhasz-Holterman MHA, Luijten JA (2000) Removal of micro-organisms by bank filtration in a gravel-sand soil. In: Jülich W, Schubew J (eds) Proceedings of the international Riverbank filtration conference. International Association of the Rhine Waterworks (IAWR), Amsterdam, Dordrecht, p 161–168

Meierhoffer R, Wegelin M (2002) Solar water disinfection, a guide for the application of SODIS. Swiss Federal Institute of Environmental Science and Technology (EAWAG), Department of Water and Sanitation in Developing Countries (SANDEC). SANDEC, Dübendorf

Menguy G et al (1980) New solar still (the wiping spherical still): design and experimention. (Malik et al. 1982)

Metcalf R (2009) The microbiology of solar water pasteurization, with applications in East Africa. Solar Cookers International. http://www.katharinehamnett.com/images/campaigns/csp_report/New-Methods-of-Water-Pasteurisation.pdf. Accessed 16 July 2009

Metge DW, Harvey RW, Aiken GR, Anders R, Lincoln G, Jasperse J (2010) Influence of organic carbon loading, sediment-associated metal oxide content distributions upon Cryptosporidium parvum removal during bank filtration operations, Sonoma County, CA. Water Res 44(4):1126–1137

Meukam P, Njomo D, Gbane A, Toure S (2004) Experimental optimization of a solar still: application to alcohol distillation. Chem Eng Process 43(12):1569–1577

Michot D, Benderitter Y, Dorigney A, Nicoullaud B, King D, Tabbagh A (2003) Spatial and temporal monitoring of soil water content with an irrigation corn crop cover using surface electrical resistivity tomography. Water Resour Res 39(5). doi:10.1029/2002WR001581

Mickley MC (2000) Major ion toxicity in membrane concentrate. American Water Works Association, Denver

Miettinen IT, Martikainen PJ, Vartiainen T (1994) Humus transformation at the bank filtration plant. Water Sci Technol 30:179–187

Miettinen IT, Vartiainen T, Martikainen PJ (1996) Bacterial enzyme activities in ground water during bank filtration of lake water. Water Res 30(10):2495–2501

MIL-STD-1629A (1980) Procedures for performing a failure mode, effects, and criticality analysis. Department of Defense, Washington, DC 20301 (AMSCN 3074)

Minasian AN, Al-Karaghouli AA (1995) An improved solar still: the wick basin type. Energy Convers Manag 36(3):213–217

Mink G, Horvath L, Evseev EG, Kudish AI (1998) Design parameters, performance testing and analysis of a double-glazed, air-blown solar still with thermal energy recycle. Sol Energy 64(4–6):265–277

Mohsen MS, Jaber JO (2001) A photovoltaic-powered system for water desalination. Desalination 138(1–3):129–136

Molnar D (1941) Hydrogeology of the catchment of the Budapest water works. Vom Wasser 15:169–180 (in German)

Morse RN, Read WRW (1968) A rational basis of engineering development of solar still. Sol Energy 12(1):5–11

Mouchot A (1869) La Chaleur Solavie et ses applications. Industrielles, Gauthier-Villars, Paris, p 1

Moustafa SMA, Brusewitz GH, Farmer DM (1979) Direct use of solar energy for water desalination. Sol Energy 22(2):141–148

Mukherjee K, Tiwari GN (1986) Economic analysis for various designs of conventional solar stills. Energy Convers Manag 26(2):155–157

Munir AKM, Rasul SB, Habibuddowla M, Alauddin M, Hussam A, Khan AH (2001) Evaluation of performance of Sono 3-Kolshi filter for arsenic removal from groundwater using zero valent iron through laboratory and field studies. Technologies for Arsenic Removal from Drinking Water, p 171–189

Murase K, Tobata H, Ishikawa M, Toyama S (2006) Experimental and numerical analysis of a tube-type networked solar still for desert technology. Desalination 190(1–3):137–146

Nafey AS, Abdelkader M, Abdelmotalip A, Mabrouk AA (2001) Solar still productivity enhancement. Energy Convers Manag 42(11):1401–1408

Naim MM, Abd El, Kawi MA (2002a) Non-conventional solar stills Part 1. Non-conventional solar stills with charcoal particles as absorber medium. Desalination 153(1–3):55–64

Naim MM, Abd El Kawi MA (2002b) Non-conventional solar stills Part 2. Non-conventional solar stills with energy storage element. Desalination 153(1–3):71–80

Namprakal P, Hirunlabh J (2007) Theoretical and experimental studies of an ethanol basin solar still. Energy 32(12):2376–2384

Nandwani SS (1990) Economic analysis of domestic solar still in the climate of Costa Rica. Int J Solar Wind Technol 7(2–3):219–227

Nandwani SS (2006) Uses of solar energy in Costa Rica. Renew Energy 31(5):689–000

Nassar YF, Yousif SA, Salem AA (2007) The second generation of the solar desalination systems. Desalination 209:177–181

National Academy of Sciences (NAS) (2008) State science and technology policy advice: issues, opportunities, and challenges. National Academies Press, Washington

National Science and Technology Council (NSTC) (1999) Renewing the federal government—University research partnership for the 21st century. NSTC Presidential Review Directive-4, Executive Office of the President, Office of Science and Technology Policy, April

Natural Resources Defense Council (NRDC) (2009) Arsenic in drinking water. http://www.nrdc.org/water/drinking/qarsenic.asp. Accessed 10 Aug 2009

Nayak JK, Tiwari GN, Sodha MS (1980) Periodic theory of solar still. Int J Energy Res 4(1):41–57

Nebbia G, Menozzi G (1966) A short history of water desalination. Acque Dolce Dal Mare, 11 Inchiesta Inter. Milano. 18–19 April 1966, p 129–172

Nestler W, Socher M, Grischek T, Schwan M (1991) River bank infiltration in the Upper Elbe River Valley—Hydrochemical aspects. IAHS Publ 202:247–356

Nijegorodov N, Jain PK, Carlsson S (1994) Thermal-electrical, high efficiency solar stills. Renew Energy 4(1):123–127

Norov EZh et al (1975) Results of tests on solar film-covered stills. Geliotekhnika 11(3/4):130–132

NSF International & Center for Disease Control and Protection (CDC) (2009) Fact sheet: Cryptosporidium and drinking water from private wells. Division of Parasitic Diseases. http://www.cdc.gov/ncidod/dpd/healthywater/factsheets/crypto.htm. Accessed 10 Aug 2009

Nuwayhid RY, Mrad F, Abu-Said R (2001) The realization of a simple solar tracking concentrator for University research applications. Renew Energy 24(2):207–222

Oltra F (1972) Saline water conversion and its stage of development in Spain. J.E.M., Madrid

Omri A (2007) Numerical investigation on optimization of a solar distiller dimensions. Desalination 206(1–3):373–379

Onyegegbu SO (1986) Nocturnal distillation in basin-type solar stills. Appl Energy 24(1):29–42

Ortiz JM, Exposito E, Gallud F, Garcia-Garcia V, Motiel V, Aldaz A (2007) Electrodialysis of brackish water powered by photovoltaic energy without batteries: direct connection behaviour. Desalination 208(1–3):89–100

Pandey GC (1982) Liquid collector efficiency solar still. Int J Energy Res 6(1):21–28

Papapetrou M, Epp C, Tzen E (2007) Autonomous desalination units based on renewable energy systems-a review of representative installations worldwide. In: Rizzuti L, Ettouney HM, Cippolina A (eds) Solar desalination for the 21st century: a review of modern technologies and researches on desalination coupled to renewable energies. Springer, Dordrecht

Parkhurst DL, Appelo CAJ (1999) User's guide to PHREEQC—A computer program for speciation, reaction-path, 1D-transport, and inverse geochemical calculations. Technical Report 99-4259, US Geol. Survey Water-Resources Investigations Report

Partinoudi V, Collins MR, Margolin A, Brannaka L (2003) Assessment of the microbial removal capabilities of riverbank filtration. Second International Riverbank Filtration conference. Proceedings, Cincinnati, OH (Microorganisms Session)

Pasteur F (1928) Compt Rend 30:187

Patel SG, Bhatnagar S, Vardia J, Ameta SC (2006) Use of photocatalysts in solar desalination. Desalination 189(1–3):287–291

Pathak N, Ghosh PK, Daga S, Shah VJ, Patel SN (2008) Animal powered mechanical device for water desalination. US Patent 7,387,728 B2

Pearce GK (2007) The case for UF/MF pretreatment to RO in seawater applications. Desalination 203:286–295

Pejack E, Al-Humaid A, Al-Dossary G, Saye R (1996) Simple devices for solar purification of water. Second World conference on Solar Cookers, Universidad Nacional, Heridia, Costa Rica, 12–15 July

PES Environmental Inc (2003) Hydrogeologic evaluation, Hanson Aggregates and Town of Windsor properties. Windsor, California, p 25 (Unpublished report)

Petela R (2003) Exergy of undiluted thermal radiation. Sol Energy 86:241–247

Peters T, Pinto D (2008) Seawater intake and pre-treatment/brine discharge—environmental issues. Desalination 221:576–584

Phillips PJ, Smith SG, Kolpin DW, Zaugg SD, Buxton HT, Furlong ET, Esposito K, Stinson B (2010a) Pharmaceutical formulation facilities as sources of opioids and other pharmaceuticals to wastewater treatment plant effluents. Environ Sci Technol (on line now). doi:10.1021/es100356f (Publication Date (Web): 4 June 2010)

Phillips PJ, Smith SG, Kolpin DW, Zaugg SD, Buxton HT, Furlong ET, Esposito K, Stinson B (2010b) Method description, quality assurance, environmental data, and other information for analysis of pharmaceuticals in wastewater-treatment-plant effluents, stream water, and reservoirs, 2004–2009. US Geological Survey Open-File Report 2010-1102, 2010

Pontius F (1994) Crafting a new arsenic rule. J Am Water Works Assoc 86(9):6–10

Porta MA, Chargoy N, Fernandez JL (1997) Extreme operating conditions in shallow solar stills. Sol Energy 61(4):279–286

Postel SL, Daily GC, Ehrlich PR (1996) Human appropriation of renewable fresh water. Science 271:785–788

Prakash J, Kavathekar AK (1986) Performance prediction of regenerative solar still. Solar Wind Technol 3(2):119–125

Prasad B, Tiwari GN (1996) Effect of glass covers inclination and parametric studies of concentrator-assisted solar distillation system. Int J Energy Res 20(6):495–505

Priscoli JD (1998) Water and civilization: using history to reframe water policy debates and to build a new ecological realism. Water Policy 1(6):623–636

Prommer H, Barry DA (2001) PHT3D—a reactive multi-component transport model for saturated porous media. Version 1.0 User's Manual, Technical Report, Contaminated Land Assessment and Remediation Research Centre, The University of Edinburgh

Pruess K (1991) TOUGH2—a general-purpose numerical simulator for multi-phase fluid and heat flow. Lawrence Berkeley Laboratory Report LBL-29400, Berkeley, CA, USA, p 102

PVK (2000) Vodarna Karany—Karany water works. Information material of the Prazske Vodovody a Kanlizace a.s., p 12
Qasim SR (1978) Treatment of domestic sewage by using solar distillation and plant culture. Environ Sci Health 13(8):615–627
Rabionovitch E (2008) Drinking with the wind. MS Thesis, Delft University of Technology, Delft, Dordrecht
Rahim NHA (2001) Utilisation of new technique to improve the efficiency of horizontal solar desalination still. Desalination 138(1–3):121–128
Rahim NHA (2003) New method to store heat energy in horizontal solar desalination still. Renew Energy 28(3):419–433
Rai SN, Tiwari GN (1982) Single basin solar still coupled with flat plate collector. Energy Convers Manag 23(3):145–149
Rail CD (1989) Groundwater contamination: sources, control, and preventative measures. Technomic Publishing Co, Lancaster
Ramadane A, Barkaoui M, Goff HL, Jurkowski R, Goff PL (1986) World Congress of Chem. Eng. III, Tokyo, 2, 532
Ray C (2008) Worldwide potential of riverbank filtration. Clean Technol Environ Policy 10(3):223–225
Ray C, Grischek T, Schubert J, Wang JZ, Speth TF (2002a) A perspective of riverbank filtration. J Am Water Works Assoc 94(4):149–160
Ray C, Melin G, Linsky RB (2002b) Riverbank filtration: improving source-water quality. Kluwer, Dordrecht
Ray C, Song TW, Lian YQ, Roadcap GS (2002c) Effect of flood-induced chemical load on filtrate quality at bank filtration sites. J Hydrol 266:235–258
Regli S, Rose JB, Haas CN, Gerba CP (1991) Modeling the risk from Giardia and viruses in drinking water. J Am Water Works Assoc 83(11):76–84
Revil A, Titov K, Doussan C, Lapenna V (2006) Applications of the self-potential method to hydrological problems. In: Vereecken H, Binley A, Cassiani G, Revil A, Titov K (eds) Applied hydrogeophysics. Springer, Dordrecht
Roberts DV (1994) Sustainable development—a challenge for the engineering profession. In: The role of engineering in sustainable development. American Association of Engineering Societies, Washington, DC
Robinson R, Ho G, Mathew K (1992) Development of a reliable low-cost reverse osmosis desalination unit for remote communities. Desalination 86:9–26
Rogers EM (1983) Diffusion of innovations, 3rd edn. Free Press, New York
Rogers EM (1995) Diffusion of innovations, 4th edn. Simon and Schuster, New York
Rommel M, Koschikowski J, Wieghaus M (2007) Solar driven desalination systems based on membrane distillation. In: Rizzuti L, Ettouney HM, Cippolina A (eds) Solar desalination for the 21st century: a review of modern technologies and researches on desalination coupled to renewable energies. Springer, Dordrecht
Rorabaugh MI (1954) Streambed percolation in development of water supplies. US Geological Survey Groundwater Notes on Hydraulics No. 25, p 13
Rosenberry DO, LaBaugh JW (2008) Field techniques for estimating water fluxes between surface water and ground water. US Geological Survey Techniques and Methods 4–D2, p 128
Rubio-Cerda E, Porta-Gandara MA, Fernandez-Zayas JL (2002) Thermal performance of the condensing covers in a triangular solar still. Renew Energy 27(2):301–308
Rubio E, Porta MA, Fernandez JL (2000) Cavity geometry influence on mass flow rate for single and double slope solar stills. Appl Therm Eng 20(12):1105–1111
Rubio E, Fernandez JL, Porta-Gandara MA (2004) Modeling thermal asymmetries in double slope solar stills. Renew Energy 29(6):895–906
Sadineni SB, Hurt R, Halford CK, Boehm RF (2008) Theory and experimental investigation of a weir-type inclined solar still. Energy 33(1):71–80
Safapouur N, Metcalf R (1999) Enhancement of water pasteurization with reflectors. Appl Environ Microb 65(2):859–861

Safrai I, Zask A (2008) Reverse osmosis desalination plants- marine environmentalist regulator point of view. Desalination 220:72–84

Sahoo BB, Sahoo N, Mahanta P, Borbora L, Kalita P, Saha UK (2008) Performance assessment of a solar still using blackened surface and thermocol insulation. Renew Energy 33(7):1703–1708

Sakthivel M, Shanmugasundaram S (2008) Effect of energy storage medium (black granite gravel) on the performace of a solar still. Int J Energy Res 32(1):68–82

Saltonstall CW (1977) Progress in membrane development for reverse osmosis. Desalination 22:229–234

Samee AM, Mirza UK, Majeed T, Ahmed N (2007) Design and performance of a simple single basin solar still. Renew Sust Energy Rev 11:543–549

SANDEC (1997) Faecal sludge quantities and characteristics. http://www.sandec.com

Sandhu C, Grischek T, Kumar P, Ray C (2010) Potential for riverbank filtration in India. Clean Technol Environ Policy (published online 5 May 2010). doi:1007/s10098-010-0298-0

Sartori E (1996) Solar still versus solar evaporator: a comparative study between their thermal behaviors. Sol Energy 56(2):199–206

Saye R, Pejack E (1994) A temperature indicator for purifying water. Proceedings of the second world conference on Solar Cookers, Universidad Nacional, Heridia, Costa Rica, 12–15 July

Schafer DC (2003) Use of aquifer testing and groundwater modeling to evaluate aquifer/river hydraulics at Louisville Water Company, Louisville, Kentucky, USA. In: Hubbs SA (ed) Riverbank filtration hydrology: impacts on system capacity and water quality. Springer, Dordrecht

Schijven JF, Medema GJ, de Nijs ACM, Elzenga JG (1996) Emission and distribution of Cryptosporidium, Giardia and enteroviruses via domestic wastewater. National Institute of Public Health and the Environment, Bilthoven, Report 289202 014 (in Dutch)

Schijven JF, Hoogenboezem W, Hassanizadeh SM et al (1999) Modeling removal of bacteriophages MS2 and PRD1 by dune recharge at Castricum, Dordrecht. Water Resour Res 35(4):1101–1111

Schijven JF, Hassanizadeh SM, de Bruin (2002) Column experiments to study nonlinear removal of bacteriophages by passage through saturated dune sand. J Contam Hydrol 58(3–4):243–259

Schmid P, Kohler M, Meierhofer R, Luzi S, Wegelin M (2008) Does the reuse of PET bottles during solar water disinfection pose a health risk due to the migration of plasticizers and other chemicals into the water? Water Res 42(20):5054–5060

Schmidt CK, Lange FT, Brauch H-J, Kühn W (2003) Experiences with riverbank filtration and infiltration in Germany. DVGW-Water Technology Center (TZW) Karlsruhe, Germany. http://www.tzw.de/pdf/bankfiltration.pdf. Accessed 15 Feb 2010

Schoenheinz D, Grischek T (2010) Behaviour of dissolved organic carbon (DOC) during bank filtration under extreme climate conditions. In: Ray C, Shamrukh M (eds) Riverbank filtration for water security in desert countries. Springer, Dordrecht

Schubert J (2002a) German experience with riverbank filtration systems. In: Ray C, Melin G, Linsky RB (eds) Riverbank filtration: improving source water quality. Kluwer, Dordrecht, p 35–48

Schubert J (2002b) Hydraulic aspects of riverbank filtration—field studies. J Hydrol 266(3): 145–161

Schwab K, Porter ME (2008) The global competitiveness report 2008–2009. World Economic Forum, Geneva, Switzerland

Schwarzer S, Vieira ME, Faber C, Muller C (2001) Solar thermal desalination system with heat recovery. Desalination 137(1):23–29

Selcuk MK (1971) Analysis, design and performance evaluation of controlled environment greenhouse. Trans ASHRAE, No. 2172

Sen A (2001) Foreword. In: Harris JM et al (eds) A survey of sustainable development—social and economic development. Island Press, Washington

Sharma L, Ray C (2010) A combined RBF and ASR system for providing drinking water in water scarce areas. In: Ray C, Shamrukh M (eds) Riverbank filtration for water security in desert countries. Springer, Dordrecht

Shawaqfeh AT, Farid MM (1995) New development in the theory of heat and mass transfer in solar stills. Sol Energy 55(6):527–535

Sheets KR, Hendrickx JMH (1995) Non-invasive soil water content measurement using electromagnetic induction. Water Resour Res 31:2401–2409
Sheffer MR, Howie JA (2003) A numerical procedure for the study of the streaming potential phenomenon in embankment dams. Proceedings symposium on the application of geophysics to engineering and environmental problems
Shetty K, Paliyath G, Pometto A, Levin RE (2006) Food Biotechnology. CRC Press, Boca Raton
Shukla SK, Sorayan VPS (2005) Thermal modeling of solar stills: an experimental validation. Renew Energy 30:683–699
Sikdar S, Glavic P, Jain RK (2004) Technological choices for sustainability. Springer, Berlin
Singh AK, Tiwari GN (1992) Performance study of double effect distillation in a multiwick solar still. Energy Conserv Manag 33(3):207–214
Singh AK, Tiwari GN, Sharma PB, Khan E (1995) Optimization of orientation for higher yield of solar still for a given location. Energy Conserv Manag 36(3):175–181
Singh HN, Tiwari GN (2004) Monthly performance of passive and active solar stills for different Indian climatic conditions. Desalination 168:145–150
Singh HN, Tiwari GN (2005) Evaluation of cloudiness/haziness factor for composite climate. Energy 30(9):1589–1601
Singh SK, Tiwari GN (1991) Analytical expression for thermal efficiency of passive solar still. Energy Convers Manag 32(6):571–576
Sodha MS, Tiwari GN (1979) Low cost portable solar still. Sunworld 3:172
Sodha MS, Kumar A, Singh U, Tiwari GN (1980a) Energy conservation of solar still. Energy Conserv 200(3):191
Sodha MS, Nayak JK, Tiwari GN, Kumar A (1980b) Double basin solar still. Energy Convers Manag 20(1):23–32
Sodha MS, Kumar A, Tiwari GN, Pandey GC (1980d) Effect of dye on thermal performance of solar still. Appl Energy 7(1–3):147–162
Sodha MS, Kumar A, Tiwari GN, Tyagi RC (1981) Simple multiple-wick solar still: analysis and performance. Sol Energy 26(2):127–131
Solar Cookers International (SCI) (2009a) Cookit. Wikia Inc. www.solarcooking.org. Accessed 2 July 2009
Solar Cookers International (SCI) (2009b) Water pasteurization indicator. Wikia Inc. http://solarcooking.wikia.com/wiki/Water_Pasteurization_Indicator. Accessed 2 July 2009
Soliman HS (1972) Effect of wind on solar distillation. Sol Energy 13(4):403–415
Soliman HS (1976) Solar still coupled with a solar water heater. Mosul University, Mosul, Iraq, p 43
Solow RM (1993) Sustainability: an economist's perspective. In: Dorfman RM, Dorfman NS (eds) Economics of the environment—selected readings. WW Norton & Company Inc, New York
Son HJ, Heo MS, Kim YG, Lee SJ (2001) Optimization of fermentation conditions for the production of bacterial cellulose by a newly isolated Acetobacter sp. A9 in shaking cultures. Biotechnol Appl Biochem 33:1–5
Srinivasan J (1993) Solar pond technology. Sadhana 18:39–55
Stevens R, Johnson R, Eckerlin H (1998) An investigation of a solar pasteurizer with an integral heat exchanger (SPIHX). American Solar Energy Society conference proceedings, 14–17 June 1998
Stonestrom DA, Constantz J (eds) (2003) Heat as a tool for studying the movement of ground water near streams. USGS Circular 1260
Stover RL (2006) Energy recovery devices for seawater reverse osmosis. EverythingAboutWater Nov:40–45
Su GW, Jasperse J, Seymour D, Constantz J (2004) Estimation of hydraulic conductivity in an alluvial system using temperatures. Ground Water 42(6):890–901
Su GW, Jasperse J, Seymour D, Constantz J (2007) Analysis of pumping-induced unsaturated regions beneath a perennial river. Water Resour Res 43:W08241
Suneja S, Tiwari GN (1998) Optimization of number of effects for higher yield from an inverted absorber solar still using Runge-Kutta method. Desalination 120(3):197–209

Suneja S, Tiwari GN (1999a) Effect of water depth on the performance of an inverted absorber double basin solar still. Energy Convers Manag 40(17):1885–1897

Swanson JC, Mueller C, Barrett S (2006) Analysis of intake and discharge salinity regimes for a desalination plant. Oceans 18–21:1–5

Swinbank WC (1963) Long-wave radiation from clear skies. Q J R Meteorol Soc 89:339–348

Talbert SG, Eibling JA, Lof GOG (1970) Manual on solar distillation of saline water. R&D Progress Report No. 546, US Department of the Interior

Tamini A (1987) Performance of a solar still with reflectors and black eye. Solar Wind Technol 4(4):443–446

Tanaka H, Nakatake Y (2005a) A simple and highly productive solar still: a vertical multiple effect diffusion-type solar still coupled with a flat-plate mirror. Desalination 173(3):287–300

Tanaka H, Nakatake Y (2005b) Factors influencing the productivity of a multiple-effect diffusion-type solar still coupled with a flat plate reflector. Desalination 186:299–310

Tanaka H, Nakatake Y (2007) Numerical analysis of the vertical multiple-effect diffusion solar still coupled with a flat plate reflector: optimum reflector angle and optimum orientation of the still at various seasons and locations. Desalination 207(1–3):167–178

Tanaka H, Nakatake Y, Watanabe K (2004) Parametric study on a vertical multiple-effect diffusion-type solar still coupled with a heat-pipe solar collector. Desalination 171:243–255

Tanaka H, Nakatane Y, Tanaka M (2005) Indoor experiments of the vertical multiple-effect diffusion-type solar still coupled with a heat-pipe solar collector. Desalination 177(1–3):291–302

Tanaka H, Nosoko T, Nagata T (2000) A highly productive basin-type-multiple-effect coupled solar still. Desalination 130(3):279–293

Tanaka H, Nosoko T, Nagata T (2002) Experimental study of basin-type, multiple-effect, diffusion-coupled solar still. Desalination 150(2):131–144

Tanaka K, Yamashita A, Watanabe K (1982) Experimental and analytical study of the tilted wick type solar still. Solar World Forum, 2. Pergamon Press, Oxford

Telkes M (1945) Solar distiller for life rafts. United States Office of Science, R&D, Report No. 5225, P.B. 21120

Teoh MM, Wang K, Bonyadi S, Chung TS (2009) Forward osmosis and membrane distillation processes for freshwater production. Innovation. The National University of Singapore & World Scientific Publishing Co., Singapore. http://www.innovationmagazine.com/innovation/volumes/v8n1/coverstory3.shtml. Accessed 10 Aug 2009

Thomson M, Miranda M, Gwillim J, Rowbottom A, Draisey I (2001) Batteryless photovoltaic reverse-osmosis desalination system. http://www.berr.gov.uk/files/file16527.pdf. Accessed 7 Feb 2009

Tiwari AK (2006) Annual performance of solar stills for different inclination of condensing covers and water Depths. PhD Thesis, IIT Delhi

Tiwari AK, Tiwari GN (2005a) An optimization of slope of condensing cover of solar still for maximum yield: in summer climatic condition. Proceedings solar world congress, Orlando, USA 6–12 Aug

Tiwari AK, Tiwari GN (2005b) Effect of condensing cover's slope on internal heat and mass transfer in distillation: an indoor simulation. Desalination 180(1–3):73–88

Tiwari AK, Tiwari GN (2006) Effect of water depth on heat and mass transfer in a passive solar still: in summer. Desalination 195(1–3):78–94

Tiwari AK, Tiwari GN (2007a) Annual performance analysis and thermal modeling of passive solar still for different inclinations of condensing cover. Int J Energy Res 31(4):1358–1382

Tiwari AK, Tiwari GN (2007b) Thermal modeling based on solar fraction and experimental study of the annual and seasonal performance of single slope passive solar still: the effect of water depths. Desalination 207:184–204

Tiwari AK, Tiwari GN (2008) Effect of inclination of condensing cover and water depth in solar still for maximum yield: in winter climatic conditions. J Solar Energy Eng 130(2):024502. doi:10.1115/1.2844450

Tiwari GN (1984) Demonstration plant of multi-wick solar still. Energy Convers Manag 24: 313–317

Tiwari GN (1992) Recent advances in solar distillation. In: Contemporary physics: solar energy and energy conservation, chapter II. Wiley Eastern Ltd., New Delhi
Tiwari GN (2002) Solar energy, fundamentals, design, modelling and applications. CRC Press, New Delhi
Tiwari GN, Bapeswar Rao VSV (1983) Transient performance of single basin solar still with water flowing over the glass cover. Desalination 48(1):101
Tiwari GN, Garg HP (1985) Studies on various designs of solar distillation systems. Solar Wind Technol 1(3):161–166
Tiwari GN, Lawrence SA (1991) Experimental evaluation of solar distiller units with FRP lining under PNG climatic conditions. Solar Energy 9:241–248
Tiwari GN, Madhuri (1987) Effect of water depth on daily yield of the still. Desalination 61:67–75
Tiwari GN, Noor MA (1996) Characterization of solar stills. Int J Sol Energy 18:147–171
Tiwari GN, Prasad B (1996) Thermal modeling of concentrator assisted solar distillation with water flow over the glass cover. Int J Sol Energy 18(3):173–190
Tiwari GN, Salem GAM (1984) Double slope FRP multi-wick solar still. Solar Wind Technol 1:229–235
Tiwari GN, Suneja S (1998) Performance evaluation of an inverted absorber solar distillation system. Energy Convers Manag 39(3):173–180
Tiwari GN, Tiwari AK (2007) Solar distillation practice for water desalination systems. Anamaya, New Delhi
Tiwari GN, Tiwari AK (2007c) Solar distillation practice for water desalination systems. Anamaya, New Delhi
Tiwari GN, Tripathi R (2003) Study of heat and mass transfer in indoor condition for distillation. Desalination 154(2):161–169
Tiwari GN, Yadav YP (1985) Economic analysis of a GI sheet multiwick solar distilation plant. Energy Convers Manag 25(4):423–425
Tiwari GN, Madhuri, Garg HP (1985) Effect of water flow over the glass cover of single basin solar still with intermitted flow of waste hot water in the basin. Energy Convers Manag 25:315–322
Tiwari GN, Gupta SP, Lawrence SA (1989) Transient analysis of solar still in the presence of dye. Energy Convers Manag 29(1):59–62
Tiwari GN, Singh HN, Tripathi R (2003a) Present status of solar distillation. Sol Energy 75: 367–373
Tiwari GN, Shukla SK, Singh IP (2003b) Computer modeling of passive/active solar stills by using inner glass temperature. Desalination 154:171–185
Tleimat BW, Howe ED (1967) Comparison of plastic and glass condensing covers for solar distillation. Sol Energy 12:293–296
Toure S, Meukam P (1997) A numerical model and experimental investigation for a solar still in climatic conditions in Abidjan. Renew Energy 11(3):319–330
Tranvik LJ, Bertilsson S (2001) Contrasting effects of solar UV radiation on dissolved organic sources for bacterial growth. Ecol Lett 4:458–463
Tripathi R, Tiwari GN (2004) Performance evaluation of a solar still by using the concept of solar fractionation. Desalination 169(1):69–80
Tripathi R, Tiwari GN (2005) Effect of water depth on internal heat and mass transfer for active solar distillation. Desalination 173(2):187–200
Tufenkji N, Miller GF, Ryan JN, Harvey RW, Elimelech M (2004) Transport of Cryptosporidium oocysts in porous media: role of straining and physicochemical filtration. Environ Sci Technol 38(22):5932–5938
Tzen E, Perrakis K, Baltas P (1998) Design of a stand alone PV-desalination system for rural areas. Desalination 119:327–334
Umarov GY, Asamov MK, Achilov BM, Sarros TK, Norov EZ, Tsagaraeva NA (1976) Modified polyethylene films for solar stills. Geliotekhnika 12(2):29–33
United Nations Children Fund (UNICEF) (2008) UNICEF handbook on water quality. New York: UNICEF. http://www.unicef.org/wash/files/WQ_Handbook_final_signed_16_April_2008.pdf
United Nations Children Fund (UNICEF) (2008) Handbook on water quality. UNICEF, New York

United Nations Educational Scientific and Cultural Organization (UNESCO) (2009) Meeting basic needs. World Water Assessment Programme (WWAP). http://www.unesco.org/water/wwap/facts_figures/basic_needs.shtml. Accessed 11 Aug 2009

United Nations General Assembly (2000) United Nations millennium declaration. Resolution A/RES/55/2, Adopted 18 Sept 2000. http://www.un.org/millennium. Accessed 30 July 2009

United States Environmental Protection Agency (USEPA) (2007) Arsenic in drinking water: basic information. http://www.epa.gov/safewater/arsenic/basicinformation.html#two. Accessed 25 July 2009

United States Environmental Protection Agency (USEPA) (2006) Basic information about E. coli 0157:H7 in drinking water. http://www.epa.gov/safewater/contaminants/ecoli.html#two. Accessed 10 Aug 2009

United States Geological Survey (USGS) (2005) USGS estimated use of water in the United States in 2000. Fact Sheet 2005-3051. http://pubs.usgs.gov/fs/2005/3051/pdf/fs2005-3051.pdf. Accessed 16 June 2009

United States Geological Survey (USGS) (2009) Emerging contaminants project. http://toxics.usgs.gov/regional/emc/

United States Geological Survey (USGS) (1969) Techniques in water-resources investigations, book 3

United States Geological Survey (USGS) (1998) Ground water and surface water: a single resource. US Geological Survey Circular 1139

United States Geological Survey (USGS) (2003) Heat as a tool for studying the movement of ground water near streams. Circular 1260

United States Environmental Protection Agency (USEPA) (1992) Consensus method for determining groundwaters under the direct influence of surface wate rusing microscopic particulate anaysis (MPA). EPA 910/9-92-029

Valsaraj P (2002) An experimental study on solar distillation in a single slope basin still by surface heating of the water mass. Renew Energy 25(4):607–612

Veerapaneni S, Long B, Freeman S, Bond R (2007) Reducing energy consumption for seawater desalination. J Am Water Works Assoc 99:95–106

Velmurugan V, Srithar K (2007) Solar stills integrated with a mini solar pond-analytical simulation and experimental validation. Deaslination 216(1–3):232–241

Verwey EJW, Overbeek JTG (1948) Theory of stability of lyophobic colloids. Elsevier, Amsterdam

Vial D, Doussau G (2003) The use of microfiltration membranes for seawater pre-treatment prior to reverse osmosis membranes. Desalination 153:141–147

Voropoulos K, Mathioulakis E, Belessiotis V (2000) Transport phenomena and dynamic modeling in greenhouse-type solar stills. Desalination 129(3):273–281

Voropoulos K, Mathioulakis E, Belessiotis V (2003) Experimental investigation of the behavior of a solar still coupled with hot water storage tank. Desalination 156(1–3):315–322

Voropoulos K, Mathioulakis E, Belessiotis V (2004) A hybrid solar desalination and water heating system. Desalination 164(2):189–195

Voutchkov N (2004) Seawater desalination costs cut through power plant co-location. Filtr Sep 41:24–26

Wade NM (2001) Distillation plant development and cost update. Desalination 136:3–12

Wang JZ, Hubbs SA, Song R (2002) Evaluation of riverbank filtration as a drinking water treatment process. AWWA Research Foundation and American Water Works Association, USA

Watanabe K (1984) Fundamental study of the solar distillation system. Research of Natural Energy 8:125

Watmuff JH, Characters WWS, Proctor D (1977) Solar and wind induced external coefficient for solar collectors. Cooperation Mediterraneenne pour l'Energie Solaire, Revue Internationale d'Heliotechnique, 2nd Quarter, p 56

Webster JG, Nordstrom DK (2003) Geothermal arsenic. In: Welch AH, Stollenwerk KG (eds) Arsenic in ground water: geochemistry and occurrence. Kluwer, Boston, p 101–126

Whillier A (1963) Plastic covers for solar collectors. Sol Energy 7(3):148–151

Whitworth TM, Siagian UW, Lee R (2003) Reverse osmosis separation of NaCl using a bentonite membrane. Sep Sci Technol 38:4009–4025

Wibulswas P, Suntrirat S, Direkstaporn B, Kiatsiriroat T (1982) Developments of solar still having acrylic plastic covers in Thailand. Alternative Energy Sources, IC, Ann Arbor Sciences, Michigan, p 385

Wibulswas P, Tadtium S (1984) Improvement of a basin type solar still by means of a vertical back wall. Int Symposium Workshop on Renewable Energy Sources, Lahore, Elvier, Amsterdam

Wiese B, Nutzmann G (2009) Transient leakance and infiltration characteristics during lake bank filtration. Ground Water 47(1):57–68

Winter TC, Harvey JW, Franke OL, Alley WM (1998) Groundwater and surface water. A single resource. US Geol Surv Circ 1139:79

World Commission on Environment and Development (1987) Our common future. Oxford University Press, Oxford

Wu YV, Sessa DJ (2004) Protein fractionation and properties of salicornia meal. J Am Oil Chem Soc 81:173–176

Yadav YP, Tiwari GN (1987) Monthly comparative performance of solar stills of various designs. Desalination 67:565–578

Yadav YP, Yadav SK (2004) Parametric studies on the transient performance of a high-temperature distillation system. Desalination 170(3):251–262

Yale University (2005) 2005 Environmental sustainability index. Yale Center for Environmental Law and Policy, Yale University; Center for International Earth Science Information Network, Columbia University; World Economic Forum, Geneva, Switzerland; Joint Research Commission, European Commission, Ispra, Italy. Yale University, New Haven, CT

Yang T, Zall RR (1984) Chitosan membranes for reverse osmosis applications. J Food Sci 49: 91–93

Yeh HM, Ten LW, Chen LC (1985) Basin-type solar distillers with operating pressure reduced for improved performance. Energy 10:683–688

Yegian DT, Andreatta D (1996) Improving the performance of a solar water pasteurizer, developments in Solar Cookers. In: Proceedings of the second world conference on Solar Cookers, Universidad Nacional, Heridia, Costa Rica, 12–15 July

Yeh H, Chen L (1986) The effects of climatic, design and operational parameters on the performance of wick-type solar distillers. Energy Convers Manag 26(2):175–180

Yensen NP (2006) Halophyte uses for the twenty-first century. In: Khan MS, Weber DJ (eds) Ecophysiology of high salinity tolerant plants. Springer, Dordrecht

Yuan G, Zhang H (2007) Mathematical modeling of a closed circulation solar desalination unit with humidification-dehumidification. Desalination 205:156–162

Zaki GM, Dali TEl, Shafieng HEl (1983) Improved performance of solar still. In: Proceedings 1st Arab international sol energy, Kuwait, p 331

Zaki GM, Al-Turki A, AI-Fatani M (1992) Experimental investigation on concentrator assisted solar stills. Sol Energy 11:193–199

Zaragoza G, Buchhloz M, Jochum P, Perez-Purra J (2007) Watergy project: towards a rational use of water in greenhouse agriculture and sustainable architecture. Desalination 211(1–3): 296–303

Zheng C (1992) MT3D version 1.8 documentation and user's guide. SS Papadopulos& Associates, Inc., Bethesda

Ziqian C, Hongfei Z, Kaiyan H, Chaochan M (2007) Steady state experimental studies on a multi-effect thermal regeneration solar desalination unit with horizontal tube falling film evaporation. Desalination 207:59–70

Zurigat YH, Abu-arabi MK (2004) Modeling and performance analysis of a regenerative solar desalination unit. Appl Therm Energy 24:1061–1072

Index

C. Ray, R. Jain (eds.), *Drinking Water Treatment,* Strategies for Sustainability,
DOI 10.1007/978-94-007-1104-4,

N

O

P

Q

R

MIX
Papier aus verantwortungsvollen Quellen
Paper from responsible sources
FSC® C105338

If you have any concerns about our products,
you can contact us on
ProductSafety@springernature.com

In case Publisher is established outside the EU,
the EU authorized representative is:
Springer Nature Customer Service Center GmbH
Europaplatz 3, 69115 Heidelberg, Germany

Printed by Libri Plureos GmbH
in Hamburg, Germany